Functional
Anatomy

To my Mother and Father
Mary Watkins 1914–2006,
William Watkins 1909–1985

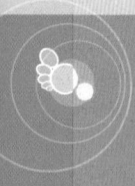

For Elsevier
Publisher: *Sarena Wolfaard*
Commissioning Editor: *Robert Edwards*
Development Editor: *Nicola Lally*
Project Manager: *Anne Dickie and Shereen Jameel*
Designer/Design Direction: *Stewart Larking*
Illustration Manager: *Bruce Hogarth*
Illustrator: *Antbits*

Functional Anatomy

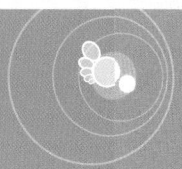

James Watkins
Professor of Biomechanics, Swansea University, UK

Series Editor
Ian Mathieson
Senior Lecturer, Wales Centre for Podiatric Studies,
University of Wales Institute, Cardiff, UK

CHURCHILL
LIVINGSTONE

ELSEVIER

Edinburgh London New York Oxford Philadelphia St Louis Sydney Toronto 2009

CHURCHILL LIVINGSTONE
ELSEVIER

First published 2009 © Elsevier Limited. All rights reserved.

ISBN 978 0 7020 3032 1

British Library Cataloguing in Publication Data
A catalogue record for this book is available from the British Library

Library of Congress Cataloging in Publication Data
A catalog record for this book is available from the Library of Congress

Notice
Neither the Publisher nor the Editors/Authors assume any responsibility for any loss or injury and/or damage to persons or property arising out of or related to any use of the material contained in this book. It is the responsibility of the treating practitioner, relying on independent expertise and knowledge of the patient, to determine the best treatment and method of application for the patient.

The Publisher

ELSEVIER your source for books, journals and multimedia in the health sciences

www.elsevierhealth.com

Working together to grow libraries in developing countries
www.elsevier.com | www.bookaid.org | www.sabre.org

ELSEVIER   BOOK AID International   Sabre Foundation

The Publisher's policy is to use **paper manufactured from sustainable forests**

Contents

Foreword

It gives me great pleasure to introduce you to *Functional Anatomy*, the first book in the *Pocket Podiatry* series. The vision of Robert Edwards, the then Commissioning Editor for Podiatry at Elsevier, the series will see volumes published over the next 5 years or so. These will build into a highly informative and clinically-oriented reference library for both under- and post-graduate podiatrists alike, as well as those from other disciplines who are involved in the management of foot, ankle and lower limb disorders. Other titles in the series include *Gait, The Ageing Foot, The Paediatric Foot, Examination and Diagnosis, Pharmacology, Footwear and Orthoses*, and *Podiatric Surgery*. It is perhaps indicative of the progress, and current status, of Podiatry that a series providing such breadth of information can be presented as reflecting the true scope of clinical practice, and as representing vital knowledge for contemporary clinical practise.

This first volume, concerned with *Functional Anatomy*, is an entirely appropriate 'series opener': podiatrists are only too aware of the importance of the weight bearing function of the foot, and its functional complexity. This volume reviews the nature of the stresses applied to the foot, their impact on various tissues, how they are managed, and how pathology can result from functional impairment. The author, James Watkins, is Professor of Biomechanics at Swansea University, but has a long history of involvement with Podiatry through a previous post in Glasgow which saw him teach the principles of mechanics to undergraduate Podiatrists for many years. This places him in a unique position to reconcile the science of biomechanics with the demands of clinical practice, and the lucid way in which this is achieved makes this volume an exciting and welcome addition to the Podiatric literature.

Professor Watkins is one of a team of writers recruited to the project who rank as some of the brightest talents working within, or with, the Podiatry profession. Each has substantial experience and a passion for their subject that is conveyed in their writing, and it is an enviable position I find myself in, being involved in the development of the manuscripts. I hope that, as more volumes are published, you judge the series to be the valuable companion to clinical podiatry it is designed to be and that your patients benefit.

Ian Mathieson

Preface

Human movement is brought about by the musculoskeletal system (bones, joints, skeletal muscles) under the control of the nervous system. The bones of the skeleton are linked together at joints in a way that allows them to move relative to each other. The skeletal muscles pull on the bones in order to control the movements of the joints and, in doing so, control the movement of the body as a whole. By coordinated activity of the various muscle groups, forces generated by the muscles are transmitted by the bones and joints to enable the individual to maintain an upright or partially upright posture, move from one place to another and manipulate objects.

The open-chain arrangement of the bones of the skeleton maximizes the range of possible body postures. However, this movement capability is only possible at the expense of low mechanical advantage of skeletal muscles, which results in relatively high forces in all components of the musculoskeletal system in most postures and movements. In response to the forces exerted on them, the musculoskeletal components experience strain (deformation). Under normal circumstances, the musculoskeletal components adapt their size, shape and structure to readily withstand the strain of everyday physical activity. This process is referred to as structural adaptation and is continuous throughout life. In the absence of disease, structural adaptation tends to maintain normal function. However, the capacity of the musculoskeletal components, especially bone, to undergo structural adaptation decreases with age. Consequently, strain that would normally result in structural adaptation in a young person may result in tissue degeneration and dysfunction in an older person.

In weightbearing activities, the function of the musculoskeletal system is to transmit the weight of the body to the ground by creating ground reaction forces at the feet to maintain upright posture and move the body in the intended direction. In walking, and other forms of locomotion such as running, hopping and jumping, the feet act alternately as shock absorbers, to cushion the impact of the foot with the ground, and propulsion mechanisms, to propel the body in the desired direction. These distinct functions are reflected in the flexible arched structure of the foot, which is indicative of the essence of functional anatomy, i.e. the intimate relationship between the structure and function of the musculoskeletal system. Accurate diagnosis and appropriate treatment of disorders of the musculoskeletal system depends largely on the clinician's knowledge and understanding of this relationship.

The purpose of this book is to develop knowledge and understanding of functional anatomy with particular reference to the foot. The book is primarily designed as a course text for undergraduate students of podiatry, but it also has a great deal to offer the practising podiatrist and other healthcare professionals, in particular, physiotherapists and occupational therapists, who deal with the acute and chronic effects of lower limb musculoskeletal pathology.

The book has seven chapters. Chapter 1 describes the elementary mechanical concepts and principles that underlie human movement, which are referred to throughout the book. Chapter 2 describes the basic structure of the body in relation to tissues, organs and systems. Chapter 3 describes the bones of the skeleton and, in particular, the features of the bones associated with force transmission and relative motion between bones. Chapter 4 describes the structure and functions of the various connective tissues, with particular reference to structural adaptation in bone. Chapter 5 describes the structure and function of the various types of joint. Chapter 6 describes the structure and function of the neuromuscular system. Chapter 7 describes the structure of the foot and, in particular, the function of the foot in walking.

No previous knowledge of functional anatomy is assumed. To aid learning, the book features a content overview at the start of each chapter, key concepts highlighted within the text, extensive use of illustrations, review questions, references to guide further reading, and an extensive glossary and index.

James Watkins

Acknowledgements

Thank you to the series editor Dr Ian Mathieson and all of the staff at Elsevier who contributed to the commissioning and production of the book. Thanks also to my academic colleagues and the large number of students who have helped me, directly and indirectly, over many years, to develop and organize the content of the book.

Elementary biomechanics

All movements and changes in movement are brought about by the action of forces. In human movements, we interact with the environment largely through pulling forces (e.g. opening a fridge door, closing a car door from the inside), pushing forces (e.g. closing a fridge door, climbing a flight of stairs), pressing forces (e.g. ringing a door-bell, pressing a key on a keyboard) or combinations of pressing forces (e.g. gripping a pen or the handle of a cup). Human movement is brought about by the neuromusculoskeletal system, i.e. the musculoskeletal system (bones, joints, skeletal muscles) under the control of the nervous system. The bones of the skeleton are linked together in a way that allows them to move relative to each other. The skeletal muscles pull on the bones in order to control the movements of the joints and, in doing so, control the movement of the body as a whole. By coordinated activity between the various muscle groups, forces generated by the muscles are transmitted by the bones and joints to enable us to maintain an upright or partially upright posture (e.g. standing, sitting), move from one place to another (e.g. crawling, walking, running, swimming) and manipulate objects (e.g. carrying a bag, lifting a box, pushing a wheelbarrow, driving a car, threading a needle).

The open-chain arrangement of the bones of the skeleton maximizes the range of possible body postures. However, this movement capability is only possible at the expense of low mechanical advantage of skeletal muscles, which results in relatively high forces in all components of the musculoskeletal system in most postures and movements. Under normal circumstances the musculoskeletal components adapt their size, shape and structure to more readily withstand the time-averaged forces exerted on them, i.e. there is an intimate relationship between the structure and function of the musculoskeletal system. To understand this relationship, it is necessary to understand elementary biomechanics. The purpose of this chapter is to develop knowledge and understanding of elementary biomechanical concepts and principles.

Force

All bodies, animate and inanimate, are continuously acted upon by forces. A force can be defined as that which alters or tends to alter a body's state of rest or type of movement. For example, in a stationary sitting position, body weight exerts a constant downward force on the body that is counteracted by upward forces exerted on the body by the seat, the floor beneath the feet and, perhaps, arm rests supporting the arms. In a standing position, the body is prevented from collapsing by the forces in the skeletal muscles that stabilize the joints. In order to start walking forward from a stationary standing position, it is necessary to push backwards against the ground; the more forcibly we push backward against the ground, the faster we move forward. Climbing stairs involves a succession of downward pushes against the stairs.

The forces that act on a body arise from interaction of the body with its environment. There are two types of interaction: contact interaction, which produces contact forces, and attraction interaction, which produces attraction forces (Watkins 2007). Contact interaction refers to physical contact between the body and its environment, such as the contact forces between our feet and the floor when standing, walking, running and jumping. Attraction interaction refers to naturally occurring forces of attraction between certain bodies that tend to make the bodies move towards each other and to maintain contact with each other after contact is made. For example, a magnetized piece of iron attracts other pieces of iron to it by the attraction force of magnetism. The human body is constantly subjected to a very considerable force of attraction, i.e. body weight, the force due to the gravitational pull of the earth. It is body weight that keeps us in contact with the ground and which brings us back to the ground should we leave it, e.g. following a jump into the air.

Mechanics and biomechanics

Forces tend to affect bodies in two ways (Watkins 2007):

- They tend to deform bodies, i.e. change the shape of the bodies by stretching, squashing, bending or twisting. For example, squeezing a tube of toothpaste changes the shape of the tube.
- They determine the movement of bodies, i.e. the forces acting on a body determine whether it moves or remains at rest and determine its speed and direction of movement if it does move.

Mechanics is the study of the forces that act on bodies and the effects of the forces on the size, shape, structure and movement of the bodies. The actual effect that a force or a combination of forces has on a body, i.e. the amount of deformation and change of movement that occurs, depends upon the size of the force in relation to the mass of the body and the mechanical properties of the body.

The mass of a body is the product of its volume and its density. The volume of a body is the amount of space that the mass occupies and its density is the concentration of matter (atoms and molecules) in the mass, i.e. the amount of mass per unit volume. The greater the concentration of mass, the larger the density. For example, the density of iron is greater than that of wood and the density of wood is greater than that of polystyrene. Similarly, with regard to the structure of the human body, bone is more dense than muscle and muscle is more dense than fat.

The mass of a body is a measure of its inertia, i.e. its reluctance to start moving if it is at rest and its reluctance to change its speed and/or direction if it is already moving. The larger the mass, the greater the inertia and, consequently, the larger the force that will be needed to move the mass or change the way it is moving. For example, the inertia of a stationary football (a small mass) is small in comparison to that of a heavy barbell (a large mass), i.e. much more force will be required to move the barbell than to move the ball.

Whereas the effect of a force on the movement of a body is largely determined by its mass, the amount of deformation that occurs is largely determined by its mechanical properties, in particular, its stiffness (the resistance of the body to deformation) and strength (the amount of force required to break the body). For a given amount of force, the higher the stiffness and the greater the strength of a body, the smaller the deformation that will occur.

Biomechanics is the study of the forces that act on and within living organisms and the effect of the forces on the size, shape, structure and movement of the organisms (Watkins 2007). In relation to humans, biomechanics is the study of the relationship between the external forces (due to body weight and physical contact with the environment) and internal forces (active forces generated by muscles and passive forces exerted on connective tissues) that act on the body and the effect of these forces on the size, shape, structure and movement of the body.

Key *Concepts*

Mechanics is the study of the forces that act on bodies and the effects of the forces on the size, shape, structure and movement of the bodies

Sub-disciplines of mechanics

The different types and effects of forces are reflected in four overlapping sub-disciplines of mechanics: mechanics of materials, fluid mechanics, statics and dynamics. Mechanics of materials is the study of the mechanical properties (strength, stiffness, resilience, toughness) of materials. Mechanics of materials includes, for example, the study of materials used to make shoes, materials used to make orthoses to be worn in shoes to treat certain foot disorders, and the effects of ageing on bone, muscle and connective tissues. Fluid mechanics is the study of: (1) the forces that affect the movement of liquids and gases, such as the flow of water in a pipe or blood flow in the cardiovascular system; and (2) the effect of liquids and gases on the movement of solids, such as the movement of the human body through water and air. Statics is the study of bodies under the action of balanced forces, i.e. study of the forces acting on bodies that are at rest or moving with constant speed in a particular direction. In these situations, the resultant force (the net effect of all the forces) acting on the body is zero. Figure 1.1A shows a man standing upright. Since the man is at rest, there are only two forces acting on him, the weight of his body W acting downward and the upward reaction force R_1 exerted by the ground. The magnitude of W and R_1 is the same, but they act in opposite directions and, therefore, cancel out, such that the resultant force acting on the man is zero.

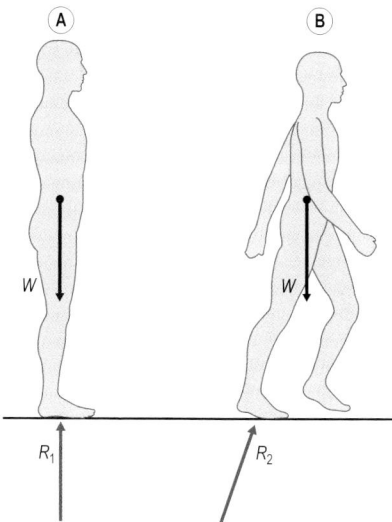

Figure 1.1 (A) The forces acting on a man standing upright and (B) just after starting to walk. W = body weight, R_1 and R_2 = ground reaction forces.

Dynamics is the study of bodies under the action of unbalanced forces, i.e. bodies moving with non-constant speed. In this situation, the result- ant force acting on the body will be greater than zero, i.e. the body will be accelerating (speed increasing) or decelerating (speed decreasing) in the direction of the resultant force. For example, as the man in **Figure 1.1A** is at rest, the resultant force acting on him will be zero. As he starts to walk (**Figure 1.1B**), he will be accelerated forward under the action of the resultant force acting on his body, i.e. the resultant of his body weight W and the force acting on his right foot R_2.

Kinematics is the branch of dynamics that describes the movement of bodies in relation to space and time (Gk. *kinema*, movement). A kin- ematic analysis describes the movement of a body in terms of distance (change in position), speed (rate of change of position) and acceler- ation (rate of change of speed). Kinetics is the branch of dynamics that describes the forces acting on bodies, i.e. the cause of the observed kinematics (Gk. *kinein*, to move).

Key *Concepts*

There are four overlapping sub-disciplines of mechanics: mechanics of materials, fluid mechanics, statics and dynamics

Forms of motion

There are two fundamental forms of motion, linear motion and angular motion. Linear motion, also referred to as translation, occurs when all parts of a body move the same distance in the same direction in the same time (Watkins 2007). In all types of self-propelled human movement, such as walking, running and swimming, the orientation of the body segments to each other continually changes and, therefore, pure linear motion seldom occurs in human movement. The human body may experience pure linear motion for brief periods in activities such as ski-jumping (Figure 1.2). When the linear movement is in a straight line, the motion is called rectilinear motion (Figure 1.2A). When the linear movement follows a curved path, the motion is referred to as curvilinear motion (Figure 1.2B).

Angular motion, also referred to as rotation, occurs when a body or part of a body, such as an arm or a leg, moves in a circle or part of

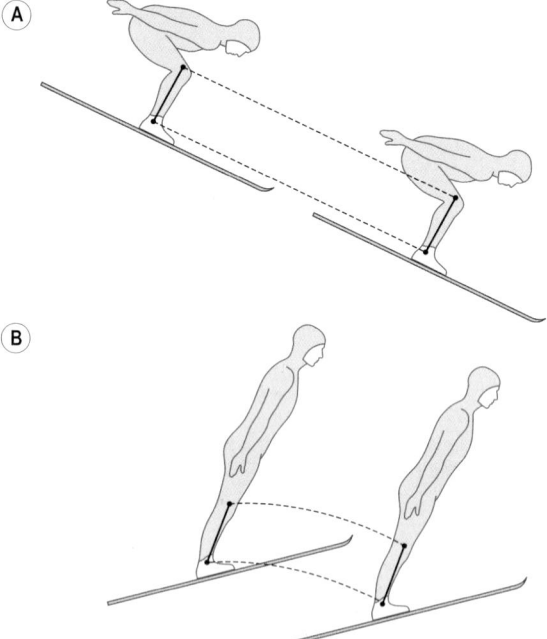

Figure 1.2 Linear motion: a ski jumper is likely to experience rectilinear motion on the runway (A) and curvilinear motion during flight (B).

a circle about a particular line in space, referred to as the axis of rotation, such that all parts of the body move through the same angle in the same direction in the same time. The axis of rotation may be stationary or it may experience linear motion (Figure 1.3). Most whole body human movements are combinations of linear and angular motion. For example, in walking, the movement of the head and trunk is fairly linear, but the movements of the arms and legs involve simultaneous linear and angular motion as the body as a whole moves forward (Figure 1.4). The movement of a multi-segmented body, like the human body, which involves simultaneous linear and angular motion of the segments, is usually referred to as general motion.

Key Concepts

There are two fundamental forms of motion, linear motion and angular motion. Most whole body human movements are combinations of linear and angular motion

Units of measurement

Commerce and scientific communication are dependent on the correct use and interpretation of units of measurement. With the advent of the industrial revolution in the 18th century and the progressive increase in international trade that resulted from it, the need for uniformity in measurement became increasingly evident. At that time, one of the most widely used systems of units was the British imperial system, but lack of clarity and consistency with regard to definitions and symbols for many variables resulted in resistance to the use of this system internationally (Rowlett 2004). The metric system of measurements originated in France around 1790. The name of the system is derived from the base unit for length, i.e. the metre, which was defined as one ten-millionth of the distance from the equator to the North Pole. In contrast to the British imperial system, each unit in the metric system has a unique definition and a unique symbol. Largely for this reason, the metric system progressively gained ground internationally. The metric system was officially adopted in the Netherlands and Luxembourg in 1820 and in France in 1837. In 1875, many of the industrialized countries signed the Treaty of the Metre, which established the International Bureau of Weights and Measures (BIPM for *Bureau International des Poids et Mesures*) and a single system of units, the International System of Units, to include all physical and chemical, metric and non-metric units. The system is usually referred to as the

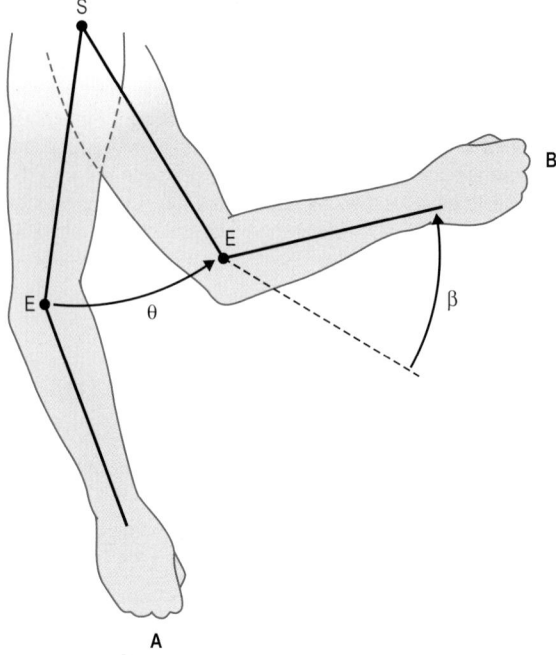

Figure 1.3 Angular motion: as the arm swings from position A to position B the upper arm rotates through an angle θ about the transverse (side-to-side) axis S through the shoulder joint and the lower arm and hand rotate through an angle β about the transverse axis E through the elbow joint.

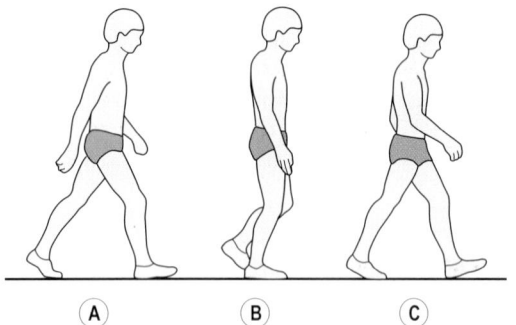

Figure 1.4 General motion: in walking, the movement of the head and trunk is fairly linear, but the movements of the arms and legs involve simultaneous linear and angular motion.

SI system after its French language name *Systeme International d'Unites*. The SI system is now the most widely used system of units, especially in science and international commerce. The system is maintained and updated by the BIPM as new units are proposed and accepted. The system now consists of a large number of units, but all of the units are derived from a set of base units. The base units for mechanical variables in the SI system are the metre (length), the kilogram (mass) and the second (time). These three units give rise to a sub-section of the SI system called the metre-kilogram-second (m-kg-s) system. The corresponding sub-section of the British imperial system is the foot-pound-second (ft-lb-s) system. These two sub-systems are shown in **Table 1.1**. Apart from in a few examples, the m-kg-s system is used in this book. In equations, a decimal point is used to indicate multiplication of one SI unit by another, e.g. N.m for newton metre.

Key *Concepts*

The International System of Units (SI system) includes all physical and chemical, metric and non-metric units. It is the most widely used system of units, especially in science and international commerce

Newton's laws of motion

Irrespective of the number of forces acting on a body, the resultant force acting on a body at rest is zero. A body at rest will only begin to move when the resultant force acting on it becomes greater than zero. Similarly, the resultant force acting on a body that is moving with uniform linear velocity, i.e. in a straight line with constant speed, is also zero. It will only change direction, accelerate or decelerate when the resultant force acting on it becomes greater than zero. Furthermore, the amount of change in speed and/or direction that occurs will depend upon the magnitude and direction of the resultant force, i.e. there is a direct relationship between change of resultant force and change in movement. Isaac Newton (1642–1725) described this relationship in what has come to be known as Newton's laws of motion.

Newton's first law of motion

To move an object from rest or to change the way it is moving, it is necessary to change the pattern of forces acting on the object. This is the basis of Newton's first law of motion.

Table 1.1 Mechanical units and (symbols) of measurement

Quantity	British imperial system	SI system
Distance	foot (ft)	metre (m)
Time	second (s)	second (s)
Speed	feet per second (ft/s)	metres per second (m/s)
Acceleration	feet per second per second (ft/s^2)	metres per second per second (m/s^2)
Mass	pound (lb)	kilogram (kg)
Linear momentum	pounds feet per second (lb.ft/s)	kilogram metres per second (kg.m/s)
Force	poundal (pdl) $1\,pdl = 1\,lb \times 1\,ft/s^2$	newton (N) $1\,N = 1\,kg \times 1\,m/s^2$
Weight*	pound force (lbf) $1\,lbf = 1\,lb \times 32.2\,ft/s^2$ $= 32.2\,pdl$	kilogram force (kgf) $1\,kgf = 1\,kg \times 9.81\,m/s^2$ $= 9.81\,N$
Pressure	pounds force per square inch (lbf/in^2)	pascal (Pa) ($Pa = N/m^2$)
Angular distance	radian (rad)	radian (rad)
Angular speed	radians per second (rad/s)	radians per second (rad/s)
Angular acceleration	radians per second per second (rad/s^2)	radians per second per second (rad/s^2)
Moment of inertia	pound foot squared ($lb.ft^2$)	kilogram metres squared ($kg.m^2$)
Angular momentum	pound foot squared per second ($lb.ft^2/s$)	kilogram metres squared per second ($kg.m^2/s$)
Turning moment	poundal foot (pdl.ft)	newton metre (N.m)
Energy and work	foot poundal (ft.pdl)	joule (J) ($J = N.m$)
Power	horsepower (hp) $1\,hp = 550\,ft\,lb/s^\dagger$	watt (W) ($W = J/s$)

*Pound force (lbf) and pound weight (lbwt) are different names for the same unit, i.e. the weight of a mass of 1 lb. Kilogram force (kgf), kilopond (kp) and kilogram weight (kgwt) are different names for the same unit, i.e. the weight of a mass of 1 kg.
†The horsepower symbol is usually written as ft lb/s, but the 'lb' is actually 'lbf' (pound force). Consequently, the correct symbol for horsepower is ft lbf/s (foot pounds force per second). Fortunately, the horsepower is rarely used in biomechanics.

An object remains at rest or continues to move with constant velocity (constant speed in a straight line) unless compelled to move or change the way it is moving by a change in the pattern of forces acting upon it.

For example, a book resting on a table will remain at rest until someone moves it. A passenger travelling in a bus moves at the same velocity as the bus. If the bus suddenly brakes the passenger will be thrown forward, especially a standing passenger, since he will tend to move forward with the velocity that the bus had immediately before braking.

Newton's second law of motion

The product of an object's mass m and linear velocity v is referred to as the linear momentum $m.v$ of the object. Any change in the velocity of an object results in a change in the linear momentum of the object. Newton's second law of motion describes the relationship between the change in linear momentum experienced by an object and the force responsible for the change.

When a force acts on an object the change in linear momentum experienced by the object takes place in the direction of the force and is proportional to the magnitude and duration of the force.

This statement can be expressed algebraically as $F.t = m.v_2 - m.v_1$, where F = magnitude of the force, t = duration of the force, m = mass of object, v_1 = velocity of object immediately prior to application of the force, and v_2 = velocity of object immediately after removal of the force. The product of force and time, $F.t$, is referred to as the impulse of the force. As $a = (v_2 - v_1)/t$, where a = the average acceleration (rate of change of velocity) of the object during the impulse $F.t$, then Newton's second law of motion can also be expressed algebraically as $F = m.a$. If v_2 was greater than v_1, i.e. the velocity of the object increased as a result of the impulse, then a is positive. If v_2 was less than v_1, i.e. the velocity of the object decreased as a result of the impulse, then a is negative. Negative acceleration is usually referred to as deceleration.

Newton's third law of motion

Objects in contact exert equal and opposite forces on each other. This is Newton's third law of motion.

When one object exerts a force on another there is an equal and opposite force exerted by the second object on the first.

Newton's law of gravitation

In addition to the three laws of motion, Newton's law of gravitation describes the naturally occurring force of attraction that is always present between any two bodies.

Every body attracts every other body with a force which varies directly with the product of the masses of the two bodies and inversely with the square of the distance between them.

Thus, the force of attraction F between two objects of masses m_1 and m_2 at a distance d apart is given by $F = (G.m_1.m_2)/d^2$ where G is a constant referred to as the Gravitational Constant and d is the distance between the centres of mass of the two bodies. In very simple terms, the law of gravitation means that the force of attraction between any two bodies will be greater the larger the masses of the bodies and the closer they are together. It is, perhaps, hard to appreciate that a force of attraction exists between any two bodies. However, the force of attraction between bodies is normally minute and has no effect on the movement of the bodies. There is, however, one body that results in a significant force of attraction between itself and other bodies, i.e. the earth. In relation to the law of gravitation, the earth is simply a massive body (radius 6.37×10^3 km; Elert 2000) with a huge mass (5.98×10^{24} kg; Elert 2000). Even though the distance between the centre of the earth (assumed to be the centre of mass of the earth) and any body on the surface of the earth (or in space close to the surface of the earth) is extremely large, the force of attraction between the earth and any other body is much larger than that which exists between any two bodies on or close to the earth's surface. This is due to the huge mass of the earth. The force of attraction between the earth and any other body is not large enough to have any effect on the movement of the earth, but it is certainly large enough to pull any body towards the earth. The force of attraction between the earth and any body is referred to as the weight of the body. This is the force of attraction that keeps us in contact with the earth or brings us back to the surface of the earth very quickly should we momentarily leave it as, for example, when jumping off the ground. By the law of gravitation, the weight W of an object of mass m may be expressed as $W = m(G.M/d^2)$ where M = mass of the earth and d = distance between the centre of the earth and the object on its surface. The term $G.M/d^2$ is usually referred to as gravity, g (not to be confused with G) which is the acceleration due to the earth's gravitational field, i.e. when an object is held above the ground and then released, it will accelerate downward at 9.81 m/s². The weight W of a mass m is usually expressed as $W = m.g$.

Units of force

In the metric system, the unit of force is the newton (N). A newton is defined as the force acting on a mass of 1 kg that accelerates it at $1 \, m/s^2$, i.e. $1 \, N = 1 \, kg \times 1 \, m/s^2$. Since the acceleration due to gravity is $9.81 \, m/s^2$ it follows that the weight of a mass of 1 kg, referred to as 1 kgf (kilogram force: see Table 1.1), is given by

$$1 \, kgf = 1 \, kg \times 9.81 \, m/s^2 = 9.81 \, N$$

A mass of 1 kg is equal to 2.2046 lb, i.e. $1 \, kgf = 2.2046 \, lbf$ (pound force). The kgf and lbf are referred to as gravitational units of force. Body weight is often recorded in kg or lb, which are units of mass. While this makes no practical difference, the correct units for weight are kgf or lbf (Watkins 2007).

Centre of gravity

The human body consists of a number of segments linked by joints. Each segment contributes to the body's total weight (Figure 1.5A). Movement of the body segments relative to each other alters the weight distribution of the body. However, in any particular body posture the body behaves (in terms of the effect of body weight on the movement of the body) as if the total weight of the body is concentrated at a single point called the centre of gravity (also referred to as centre of mass) (Figure 1.5B). Body weight acts vertically downward from the centre of gravity along a line called the line of action of body weight. The concept of centre of gravity applies to all bodies, animate and inanimate.

Key Concepts

The centre of gravity of an object is the point at which the whole weight of the object can be considered to act

The position of an object's centre of gravity depends on the distribution of the weight of the object. For a regular-shaped object with uniform density such as a cube, oblong or sphere, the centre of gravity is located at the object's geometric centre. However, if the object has an irregular shape or non-uniform density, like the human body, the position of the centre of gravity will reflect the mass distribution and it may be inside or

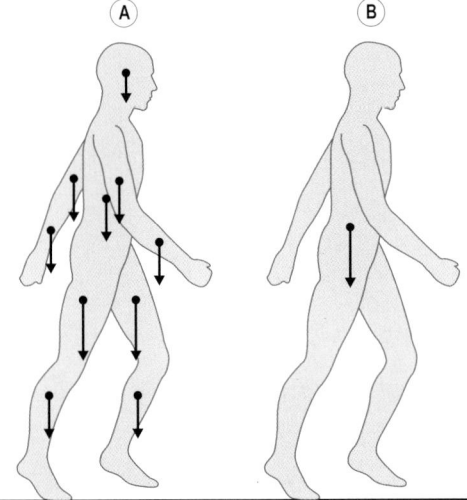

Figure 1.5 (A) Centres of gravity of body segments and lines of action of the weights of the segments. (B) Centre of gravity of the whole body and line of action of body weight.

outside the body (**Figure 1.6**). Movements that involve continuous change in the orientation of the body segments to each other, such as walking and running, result in continuous change in the position of the centre of gravity.

Stability

Figure 1.7 shows a regular cube-shaped block of wood resting on a horizontal surface. The centre of gravity of the block of wood is located at its geometric centre and the line of action of the weight of the wood intersects the base of support ABCD on which it is resting. If the block of wood is tilted over on any of the edges of the base of support, AB, BC, CD or AD, it will return to its original position provided that, at release, the line of action of its weight intersects the plane of the original base of support ABCD. This situation is shown in **Figure. 1.7C** with respect to the edge BC. However, if, at release, the line of action of its weight does not intersect the original base of support, the block of wood will fall onto one of its other faces as shown in **Figure 1.7D and E**. With respect to a particular base of support, an object is stable when the line of action of its weight intersects the plane of the base of support and unstable when it does not. Consequently, the block of wood in **Figure 1.7** is stable with

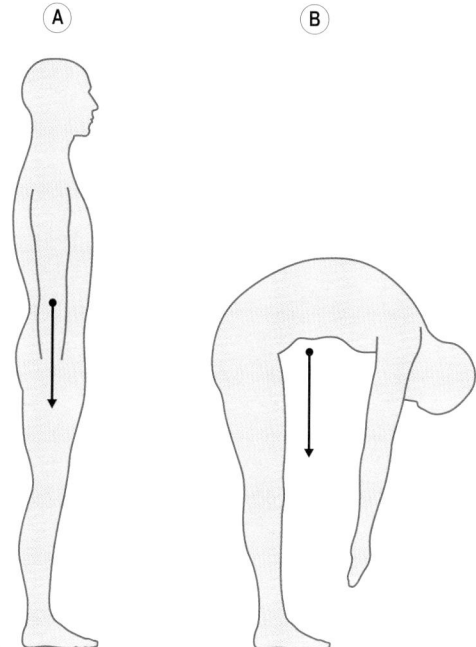

Figure 1.6 Position of the whole body centre of gravity when standing upright (A) and bending forward (B).

respect to the base of support ABCD in the positions shown in **Figure 1.7B and C**, and unstable with respect to the base of support ABCD in the position shown in **Figure 1.7D**.

Key *Concepts*

> With respect to a particular base of support, an object is stable when the line of action of its weight intersects the plane of the base of support and unstable when it does not

With regard to human movement, the terms stability and balance are often used synonymously. Maintaining stability of the human body is a fairly complex, albeit largely unconscious, process (Roberts 1995). When standing upright the line of action of body weight intersects the base of support formed by the area beneath and between the feet (**Figure 1.8A and B**).

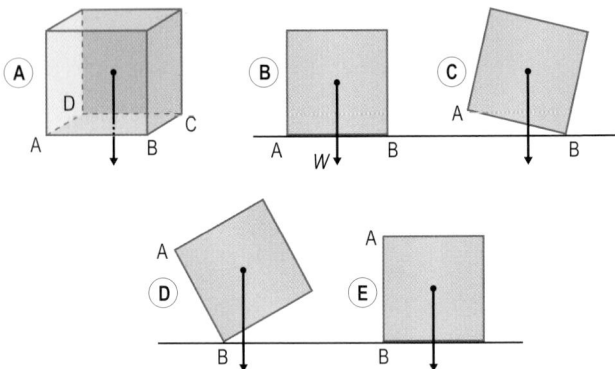

Figure 1.7 The line of action of the weight of a cube in relation to its base of support.

The size of the base of support can be increased by moving the feet further apart. For example, moving one foot to the side increases side-to-side stability (Figure 1.8C) and moving one foot in front of the other increases anteroposterior stability (Figure 1.8D).

In general, the lower the centre of gravity and the larger the area of the base of support, the greater the stability. The recumbent position is the most stable position of the human body since it is the position in which the area of the base of support is greatest and the height of the centre of gravity lowest. The degree of muscular effort needed to maintain stability tends to decrease as the area of the base of support increases. For example, it is usually easier, in terms of muscular effort, to maintain stability when standing on both feet than when standing on one foot. Similarly, it is usually less tiring to sit than to stand, and less tiring to lie down than to sit. A person recovering from a leg injury may use crutches or a walking stick in order to relieve the load on the injured limb. The use of crutches or a walking stick also increases the area of the base of support and makes it easier for the user to maintain stability (Figure 1.8E and F).

Movement of the body from one base of support to another, such as in moving from standing to sitting, illustrates the unconscious way in which the balance systems of the body automatically redistribute body weight to maintain stability. Figure 1.9 shows a person moving from a standing position to sitting on a chair. The person moves his feet close to the front of the chair and then lowers his body by flexing his knees and bending his trunk forward while maintaining the same base of support, i.e. the area beneath and between his feet (Figure 1.9A and B). He may or may not take hold of the sides of the chair as his thighs approach the seat of the chair. If he does take hold of the chair, his base of support

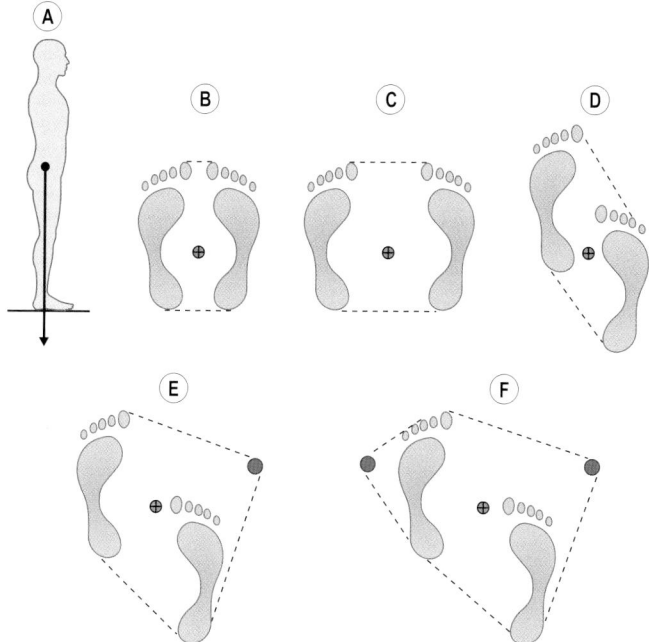

Figure 1.8 The line of action of body weight in relation to the base of support. (A, B) Standing upright. (C) Standing upright with feet apart, side-by-side. (D) Standing upright with the left foot in front of the right foot. (E) Standing with the aid of a walking stick in the right hand. (F) Standing with the aid of crutches or two walking sticks. The symbol ⊕ denotes the point of intersection of the line of action of body weight with the base of support.

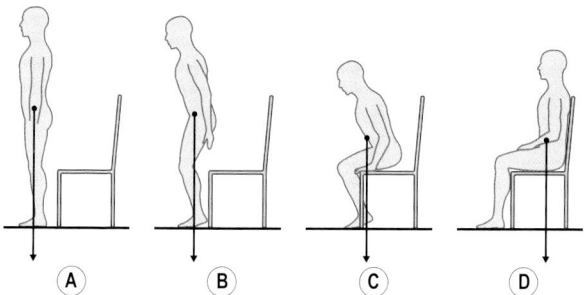

Figure 1.9 The line of action of body weight in relation to the base of support when moving from standing to sitting.

immediately increases to include the area bounded by the legs of the chair as well as that beneath and between his feet, but the line of action of his body weight will still be over the area between his feet (Figure 1.9C). When his thighs come close to the seat of the chair he begins to transfer his weight from over his feet to over the seat by gently rocking the trunk backward (Figure 1.9D). These movements are reversed when moving from a sitting to a standing position.

Load, strain and stress

A load is any force or combination of forces applied to an object (Watkins 2007). There are three types of load: tension, compression and shear (Figure 1.10). Loads tend to deform the objects on which they act. Tension is a pulling (stretching) load that tends to make an object longer and

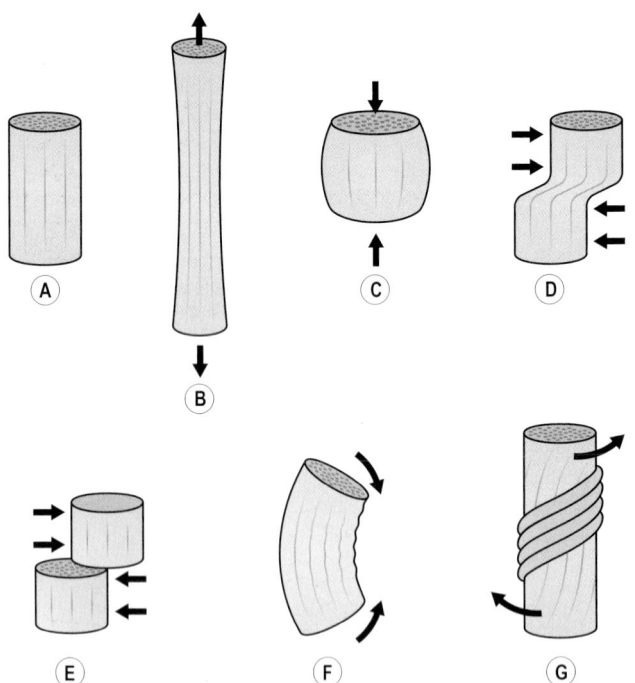

Figure 1.10 Types of load: (A) unloaded, (B) tension, (C) compression, (D) shear, (E) shear producing friction, (F) bending, (G) torsion.

thinner along the line of the force (Figure 1.10A and B). Compression is a pushing or pressing load that tends to make an object shorter and thicker along the line of the force (Figure 1.10A and C). A shear load is comprised of two equal (in magnitude), opposite (in direction), parallel forces that tend to displace one part of an object with respect to an adjacent part along a plane parallel to and between the lines of force (Figure 1.10A and D). The cutting load produced by scissors and garden shears is a shear load, while the cutting load produced by a knife is a compression load. It is also a shear load that forces one object to slide on another (Figure 1.10E). The sliding or tendency to slide is resisted by a force called friction, which is exerted between and parallel to the two contacting surfaces.

The three types of load frequently occur in combination, especially in bending and torsion (Figure 1.10A, F and G). An object subjected to bending experiences tension on one side and compression on the other. An object subjected to torsion simultaneously experiences tension, compression and shear.

In mechanics, the deformation of an object that occurs in response to a load is referred to as strain. For example, when a muscle contracts it exerts a tension load on the tendons at each end of the muscle and, consequently, the tendons experience tension strain, i.e. they are very slightly stretched. Similarly, an object subjected to a compression load experiences compression strain and an object subjected to a shear load experiences shear strain. Strain denotes deformation of the intermolecular bonds that comprise the structure of an object. When an object experiences strain, the intermolecular bonds exert forces that tend to restore the original (unloaded) size and shape of the object. The forces exerted by the intermolecular bonds of an object under strain are referred to as stress. Stress is the resistance of the intermolecular bonds to the strain caused by the load.

The stress on an object resulting from a particular load is distributed throughout the whole of the material sustaining the load. However, the level of stress in different regions of the material varies depending upon the amount of material sustaining the load in the different regions; the more material sustaining the load, the lower the stress. Consequently, stress is measured in terms of the average load on the plane of material sustaining the load at the point of interest.

Key Concepts

A load is any force or combination of forces applied to an object. Strain is the deformation of an object that occurs in response to a load. Stress is the resistance of the intermolecular bonds to the strain caused by the load

Tension stress

Figure 1.11A shows a person standing upright with the line of action of body weight slightly in front of the ankle joints. In this posture stability is maintained by isometric (static) contraction of the ankle plantar flexors as shown in the simple two-segment model in **Figure 1.11B**. If the force exerted by the ankle plantar flexors is 350 N (in each leg) and the cross-sectional area of the Achilles tendon at P in **Figure 1.11B**, perpendicular to the tension load, is 1.8 cm^2, then the tension stress on the tendon at P is 194.4 N/cm^2 , i.e.

$$\text{tension stress at P} = \frac{350 \, \text{N}}{1.8 \, \text{cm}^2}$$
$$= 194.4 \, \text{N/cm}^2 \text{ (newtons per square centimetre)}$$

In the SI system, the unit of stress is the pascal (Pa), which is defined as the stress produced by a force of one newton uniformly distributed over an area of one square metre (1 Pa = 1 N/m^2). As 1 N/cm^2 = 10 000 Pa,

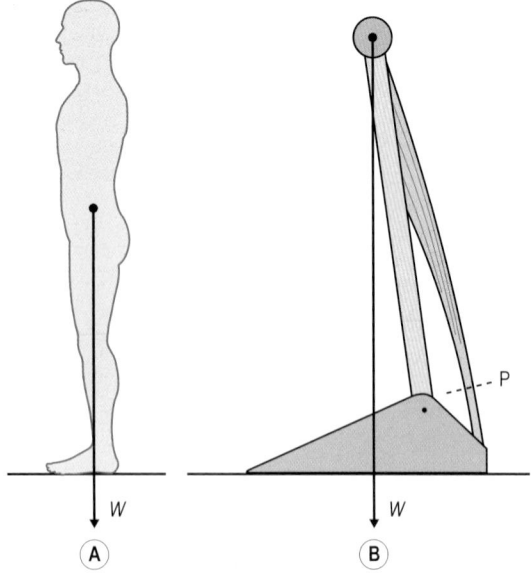

Figure 1.11 Tension load on the Achilles tendon.

the tension stress on the Achilles tendon is equivalent to 1 944 000 Pa or 1.944 MPa (MPa = megapascal = 10^6 Pa).

Compression stress

When standing barefoot, as in **Figure 1.12A**, the ground reaction force exerts a compression load on the contact area of the feet. In an adult, the contact area is approximately 260 cm^2 (both feet) (Hennig et al 1994). For a person weighing 687 N (70 kgf), the compression stress on the contact area of the feet (on a level floor, contact area perpendicular to the compression load) is 2.64 N/cm^2, i.e.

$$\text{compression stress} = \frac{687\,\text{N}}{260\,\text{cm}^2}$$
$$= 2.64\,\text{N/cm}^2 = 26\ 400\ \text{Pa}$$
$$= 26.4\,\text{kPa (kPa = kilopascal} = 10^3\ \text{Pa)}$$

By raising the heels off the ground the contact area is approximately halved (**Figure 1.12B**). Since the compression load (body weight) is the same as before, it follows that the compression stress on the reduced contact area is approximately doubled. Compression stress is usually referred to as pressure.

   In some sports played on grass pitches, such as soccer and rugby, the players wear studded boots to reduce the risk of slipping. Ideally, the studs should sink fully into the playing surface when weight-bearing;

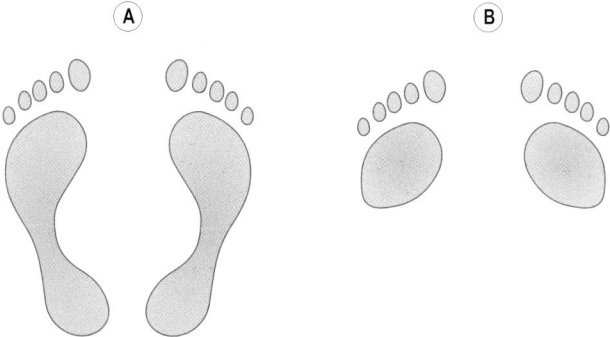

Figure 1.12 Supporting area of the feet. (A) Normal upright standing posture, (B) standing upright with the heels off the floor.

Figure 1.13 Effect of hardness of playing surface on the distribution of the ground reaction force on the sole of a studded boot. (A) Studs sink fully into the surface on a soft pitch, (B) studs do not sink fully into the surface on a hard pitch.

this would reduce the risk of slipping and ensure that the ground reaction force is distributed evenly across the whole of the soles of the boots (Figure 1.13A). However, when playing on a hard surface, the studs may not sink fully into the playing surface such that the ground reaction force will be transmitted directly by the studs and only indirectly by the soles of the boots (Figure 1.13B). The combined area of the studs is very small compared to the soles of the boots. Consequently, there will be an increase in pressure on those parts of the feet directly above the studs. The actual pressure on any part of each foot will depend upon the flexibility of the soles of the boots; the more flexible the soles, the greater the increase in pressure above the studs. Any kind of propulsive or braking movement (starting, stopping, turning) will increase the pressure even more. As the pressure increases, so does the risk of injury to the feet. There are 26 bones and about 40 joints in each foot. Consequently, most movements of the foot involve a large number of joints. Boots with inflexible soles will seriously impair the natural movement of the feet and probably result in blisters, calluses and other disorders.

Shear stress

Many of the joints, especially those in the lower back and pelvis, are subjected to shear load during normal everyday activities such as standing and walking. For example, in walking, there is a phase when one leg supports the body while the other leg swings forward (Figure 1.14). In this situation, the unsupported side of the body tends to move downward relative to the supported side subjecting the pubic symphysis joint to shear load. The area of the pubic symphysis in the plane of the shear load is approximately $2\,cm^2$. If the shear load at the instant shown

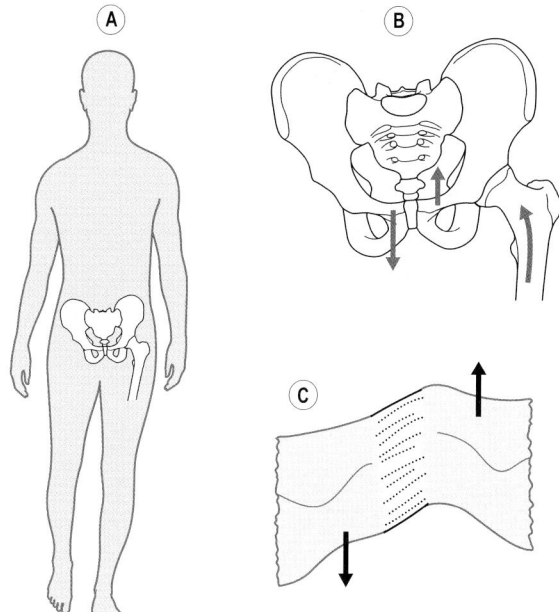

Figure 1.14 Shear load on the pubic symphysis resulting from single leg support while walking.

in Figure 1.14 is, for example, 20 N, then the shear stress on the joint is 10 N/cm², i.e.

$$\text{shear stress} = \frac{20\,\text{N}}{2\,\text{cm}^2}$$
$$= 10\,\text{N/cm}^2 = 100\,000\,\text{Pa} = 100\,\text{kPa}$$

Friction

When one object moves or tends to move across the surface of another, there will be a force parallel to the surfaces in contact that will oppose the movement or tendency to move. This force is called friction. Consider a block of wood resting on a level table (Figure 1.15). The only forces acting on the block are the weight of the block W and the force R exerted by the table on the block. Since the block is at rest, R is equal and opposite to W (Figure 1.15A). If an attempt is made to push the block along the surface of the table by applying a horizontal force P, the frictional force F between the contacting surfaces will begin to operate and oppose the tendency of the block to move (Figure 1.15B). As P increases, so does F until the block begins to move, i.e. F has a maximum value. This value is directly proportional to the degree of roughness of the two surfaces in contact and the force R. The three variables are related as follows:

$$F = \mu.R$$ **Eq. 1.1**

where μ (Greek letter mu) is a measure of the roughness of the two surfaces in contact and is called the coefficient of friction between the two surfaces. The force R is the normal reaction force, i.e. the component of the force exerted between the two surfaces that is perpendicular to the plane of contact between the two surfaces.

The magnitude of μ depends upon the types of surface in contact and whether the surfaces are sliding on each other. Surfaces are never perfectly smooth and the minor irregularities of the contacting surfaces

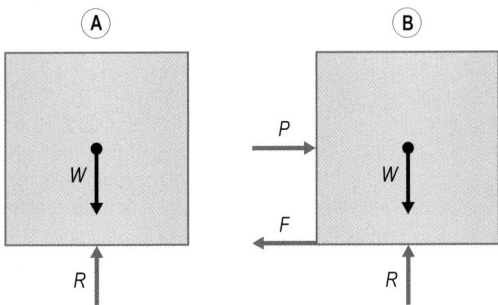

Figure 1.15 (A) The forces acting on a block of wood resting on a level table. (B) The forces acting on a block of wood resting on a level table, but tending to slide horizontally. W = weight of the block, R = normal reaction force, P = horizontal force applied to the side of the block, F = friction.

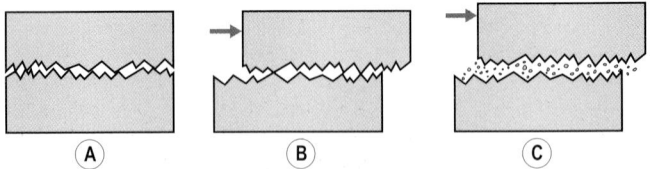

Figure 1.16 (A) Interdigitation of surface irregularities. (B) Slight separation of surfaces as a result of sliding. (C) Complete separation of surfaces as a result of lubrication.

interdigitate and resist sliding between the surfaces (Figure 1.16A). To initiate sliding, the minor irregularities have to be dragged over each other. In doing so, the surfaces tend to separate slightly, which reduces the resistance to sliding, providing that the sliding is maintained (Figure 1.16B). Consequently, for any two surfaces in contact, μ (and therefore F) will be slightly less when the surfaces are sliding on each other than when the surfaces are not sliding on each other but tending to slide. Therefore, for any two surfaces, there is a coefficient of limiting (static) friction μ_L and a coefficient of sliding (dynamic) friction μ_S.

When no sliding occurs, the normal reaction force is distributed over those parts of the adjacent surfaces that are in contact; this will include the irregularities and some of the surfaces between the irregularities (Figure 1.16A). However, during sliding, the normal reaction force is exerted almost entirely by the irregularities (Figure 1.16B). Consequently, during sliding, the pressure exerted by the irregularities of one surface on the other surface is likely to be considerably increased and result in wear of the surface (similar to the effect of sandpaper on wood). The introduction of a fluid between the surfaces tends to separate not only the surfaces, but also the surface irregularities resulting in a considerable reduction in friction and wear. This is the principle of lubrication (Figure 1.16C). In the absence of lubrication, μ is about 0.25–0.50 for wood on wood, 0.15–0.60 for metal on metal, 0.20–0.60 for wood on metal and 0.4–0.9 for wood on rubber (Serway & Jewett 2004).

Equation 1.1 shows that the amount of friction, limiting and sliding, between two surfaces is independent of the area of contact when the normal reaction force stays the same. The compression stress on the surfaces will vary with the area of contact, but the amount of friction will not change if the normal reaction force remains constant.

The development of adequate friction between the human body and the environment is essential for most actions of daily living, in particular, body transport (friction between the feet and the floor) and manipulation of objects (friction between the fingers and objects). The importance of the need for adequate friction in these types of action is, perhaps, more obvious in sport where the quality of performance is likely to depend largely upon the ability of the players to create adequate friction between feet and playing surface and between hands and racket or other implement. In such cases, adequate friction is maintained by using materials that have appropriate coefficients of friction with the playing surface and with the hands. For example, in volleyball, basketball, squash and badminton, the soles of shoes are normally made of materials that will provide adequate friction with the playing surface. It follows that for most indoor sports, the playing surfaces should not be highly polished since this will reduce the coefficient of friction and increase the possibility of slipping. However, too much friction is likely to impair performance and result in injury, especially in sports that require rapid changes in speed and/or direction. Ideally, the sole of the shoe should turn as the player turns, but excessive friction may prevent the shoe turning and result in a twisting injury to the ankle or knee. This is also a potential problem in sports played on grass pitches where the players use studded boots. In these situations, the horizontal forces produced between boots and playing surface are largely shear forces on the studs rather than friction. However, if the studs are too long and sink fully into the pitch, the sole of the boot may not turn as the player turns, which will increase the risk of injury.

There are many non-sporting situations in which it is important to ensure adequate friction in order to reduce the risk of injury. For example, an injured or aged person may rely heavily on walking sticks or crutches for support. It is very important that the sticks or crutches do not slip on the floor. Rubber has a high coefficient of friction with most materials and, consequently, rubber tips are usually fitted to the ends of the walking sticks and crutches to reduce the risk of slipping.

Whereas the development of adequate friction between the human body and the environment is essential for daily living, friction between the different body tissues inside the body must be reduced as much as possible in order to minimize the risk of injury or wear. The human body is comprised of a number of different tissues that lie adjacent to each other. Even the slightest movement involves a certain amount of sliding of the various tissues on each other and, consequently, a certain amount of friction between the tissues. Whenever friction develops, a certain amount of heat is generated. Too much heat will injure or wear body tissues. Minimizing friction is particularly important in the major weight-bearing joints of the body, i.e. the hips, knees and ankles. The articular surfaces of these joints are under considerable pressure even when the person is just standing upright and any kind of propulsive movement of the legs will increase the pressure even more. The greater the pressure, the greater the friction between the articular surfaces.

Wearing of articular surfaces is similar to the wearing of brake pads in the wheels of a motor vehicle. When braking occurs, an enormous amount of friction and, therefore, heat is generated between brake pad and wheel. Consequently, the brake pads eventually wear out and have to be replaced. In a healthy joint, any wear is usually repaired by normal metabolic processes. However, progressive joint degeneration will occur when the rate of wear outpaces the rate of repair.

In machines, parts that slide on each other are usually highly polished and friction is reduced even more by lubricating the sliding surfaces with oil or grease. Similar mechanisms exist within the human body. All the freely moveable joints of the body are lined by synovial membrane that produces synovial fluid (**Figure 1.17**). The latter is a transparent, viscous fluid, resembling the white of an egg, that lubricates and nourishes the articular surfaces (articular cartilage) of the joints. The articular surfaces are normally extremely smooth so that in association with the synovial fluid $\mu_L \simeq 0.01$ and $\mu_S \simeq 0.003$ (Serway & Jewett 2004). Consequently, the amount of friction developed in healthy joints during normal movements is usually extremely small.

The very low level of friction between the articular surfaces in a healthy synovial joint is due to the viscosity (slipperiness) of the synovial fluid. The viscosity of synovial fluid is largely due to its concentration of hyaluronic acid. With age, the concentration of hyaluronic acid tends to decrease which, in turn, decreases the viscosity of the synovial fluid (Divine et al 2007). The decrease in viscosity increases μ_L and μ_S, which, in turn, increases the friction between the articular surfaces. The increased friction results in a progressive degeneration of the articular surfaces and, ultimately, osteoarthritis. Osteoarthritic joints are characterized by synovial fluid that contains 30%–50% less hyaluronic acid than healthy joints (eOrthopod 2008). One form of treatment for osteoarthritis, referred to as viscosupplementation, involves artificially increasing the concentration of hyaluronic acid in synovial fluid by injecting it into the cavity of the affected joints.

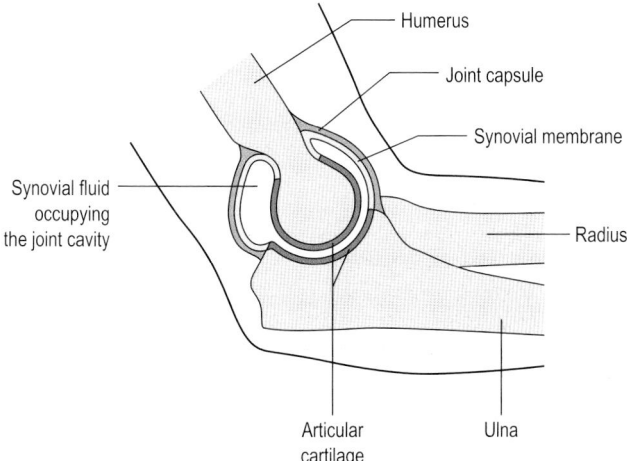

Figure 1.17 Vertical section through the elbow joint.

As part of a treatment programme, viscosupplementation has been shown to reduce joint pain in some patients (Bagga et al 2006; Divine et al 2007).

As well as in joints, synovial membranes are located in other parts of the body where different tissues slide over each other. For example, most of the tendons of muscles that cross the wrists and ankles pass over, through or around bones and ligaments. To reduce friction between the tendons and adjacent structures the tendons are enclosed within synovial sheaths (**Figure 1.18**). A synovial sheath consists of a closed

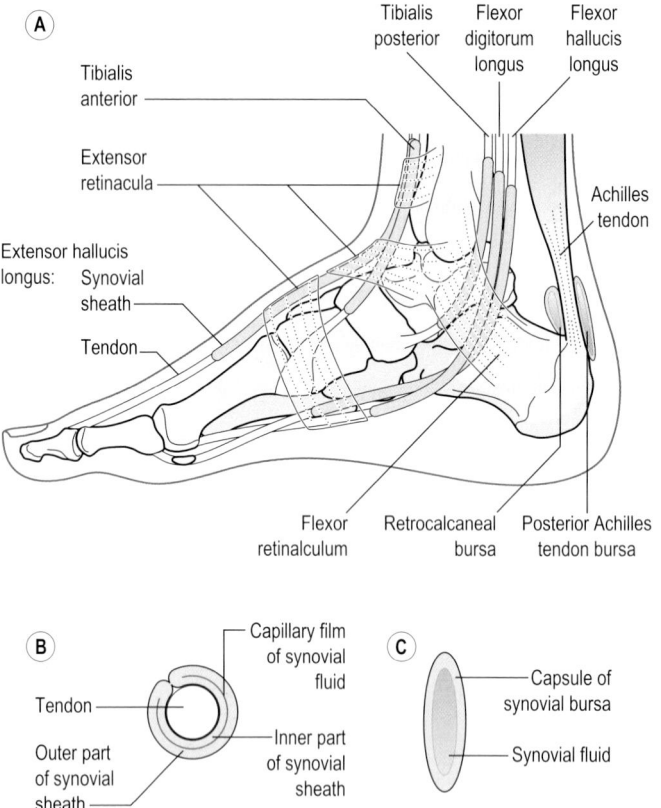

Figure 1.18 (A) Synovial sheaths, bursae and retinacula on the medial aspect of the ankle and foot. (B) Cross section through a tendon and associated synovial sheath. (C) Cross section through a bursa.

flattened sac comprised of synovial membrane containing a capillary film of synovial fluid. The synovial sheath forms a protective sleeve around the tendon. Unaccustomed overuse of the associated muscles is likely to result in tenosynovitis, i.e. a condition characterized by inflammation of the synovial sheaths. A synovial bursa is a closed sac comprised of synovial membrane containing synovial fluid that is interposed between different tissues that slide on each other. Bursae are located most frequently between the deeper layers of the skin and underlying bone (subcutaneous bursa), between tendons and bone (subtendinous bursa) and between individual muscles (submuscular bursa). Unaccustomed pressure on a bursa is likely to result in bursitis, i.e. a condition characterized by inflammation of the bursa. Normally, there is a subtendinous bursa between the Achilles tendon and calcaneus; this is referred to as the retrocalcaneal bursa or anterior Achilles tendon bursa (**Figure 1.18**).

Key *Concepts*

The viscosity of synovial fluid is largely due to its concentration of hyaluronic acid. With age, the concentration of hyaluronic acid tends to decrease which, in turn, decreases the viscosity of the synovial fluid and accelerates the process of osteoarthritis. Viscosupplementation is a form of treatment for osteoarthritis in which hyaluronic acid is injected into the joint cavity of the affected joint in order to artificially increase the concentration of hyaluronic acid

A blister that forms on the skin functions rather like a bursa and occurs in response to unaccustomed friction and/or pressure on the skin that may arise, for example, from ill-fitting shoes rubbing against the feet or the handle of a screwdriver rubbing against the hand. The body responds to this unaccustomed increase in friction and/or pressure by producing a layer of fluid between the superficial and deep layers of the skin, thereby protecting the deep layers from further damage. In the long-term, the body will respond to a sustained increase in friction and/or pressure on a particular part of the skin by thickening the superficial layer of the skin and/or producing a bursa. For example, in comparison with other parts of the body, the skin on the inferior surface of the heel and ball of each foot is subjected to fairly sustained pressure and/or frictional force. Not surprisingly, the skin on the heel and ball of each foot is usually much thicker than in other places, except, perhaps, for the palmer surfaces of the hands of manual workers. Similarly, a bursa may form, especially in response to a prolonged increase in friction. For example, a bursa may form between the Achilles tendon and the skin; this is referred to as the posterior Achilles tendon bursa (**Figure** 1.18).

Key *Concepts*

Whereas the development of adequate friction between the human body and the environment is essential for daily living, friction between the different body tissues inside the body must be reduced as much as possible in order to minimize the risk of injury or wear

---

Musculoskeletal system function

Posture refers to the orientation of body segments to each other and is usually applied to static or quasi-static positions such as sitting and standing. When standing upright there are two forces acting on the body, body weight and the ground reaction force (**Figure** 1.19A). The ground reaction force is the force exerted by the ground on the body. When standing upright the ground reaction force is equal in magnitude but opposite in direction to body weight. The combined effect of body weight and the ground reaction force is a compression load that tends to collapse the body in a heap on the ground. This compression load increases with any additional weight carried by the body (**Figure** 1.19B). To prevent the body from collapsing while simultaneously bringing about desired movements, the movements of the various joints need to be carefully controlled by coordinated activity between the various muscle groups. For example, when standing upright the joints of the neck, trunk and legs must be stabilized by the muscles that control them, other-wise the body would collapse (**Figure** 1.20). Consequently, the weight of the whole body is transmitted to the floor by the feet, but the weights of the individual body segments above the feet (head, arms, trunk and legs) are transmitted indirectly to the floor by the skeletal chain formed by the bones and joints of the neck, trunk and legs.

Transmitting body weight to the ground while maintaining an upright body posture illustrates the essential feature of musculoskeletal function, i.e. the generation (by the muscles) and transmission (by the bones and joints) of forces. In biomechanical analysis of human movement, the forces generated and transmitted by the musculoskeletal system are referred to as internal forces, and forces that act on the body from external sources, such as body weight, ground reaction force, water

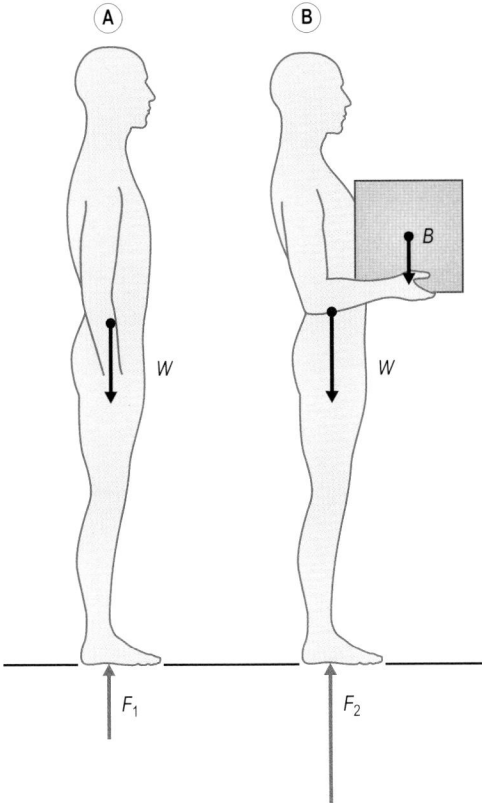

Figure 1.19 Compression load on the body in upright postures. W = body weight; B = weight of object; F_1, F_2 = ground reaction forces.

resistance and air resistance, are referred to as external forces. The musculoskeletal system generates and transmits internal forces to counteract the effects of gravity and create the ground reaction forces (and propulsion forces in water) necessary to maintain upright posture, transport the body and manipulate objects, often simultaneously (Watkins 2007).

Key *Concepts*

The function of the musculoskeletal system is to generate and transmit forces to enable us to maintain an upright or partially upright posture, move from one place to another and manipulate objects

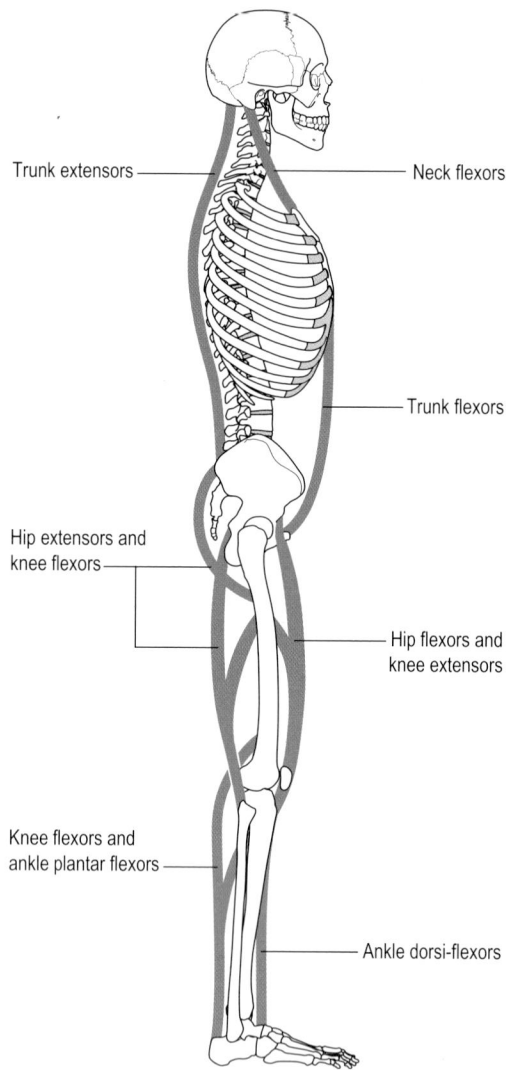

Figure 1.20 Location of the main muscle groups responsible for maintaining standing posture.

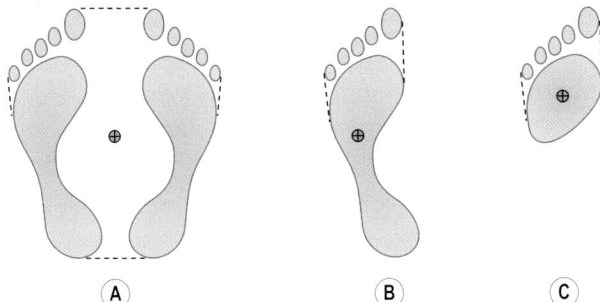

Figure 1.21 Centre of pressure ⊕ standing upright on both feet (A), standing on the left foot (B) and standing on the left foot with the heel off the floor (C).

Centre of pressure

The ground reaction force is distributed across the whole of the area of contact between the feet and the floor. Figure 1.21A shows the contact area when standing barefoot on both feet; the contact area is much smaller than the area of the base of support. Figure 1.21B shows the contact area when standing on one foot and Figure 1.21C shows the contact area when standing on one foot with the heel raised off the ground. In Figures. 1.21B and C the contact area is very similar to the base of support. Whereas the ground reaction force is distributed across the whole of the contact area, the effect of the ground reaction force on the movement of the body is as if the ground reaction force acts at a single point, which is referred to as the centre of pressure (just as the whole weight of the body appears to act at the whole body centre of gravity in terms of the effect of body weight on the movement of the body). If we could stand upright perfectly still, the centre of pressure would coincide with the point of intersection of the line of action of body weight with the base of support. However, it is very difficult to stand perfectly still; this is reflected in a certain amount of sway (forward–backward and sideways movement of the centre of gravity). The balance mechanisms are constantly active to control sway and maintain upright posture.

Vector and scalar quantities

All quantities in the physical and life sciences can be categorized as either scalar or vector quantities. Quantities that can be completely specified by their magnitude (size) are called scalar quantities. These include volume, area, time, temperature, mass, distance and speed. Quantities that require specification in both magnitude and direction are called vector quantities. These include displacement (distance in a given direction), velocity (speed in a given direction), acceleration and force. In a vector diagram, each vector is represented by an arrow; the length of the arrow, with respect to an appropriate scale, corresponds to the magnitude of the vector and the orientation of the arrow, with respect to an appropriate reference axis (usually horizontal or vertical) indicates the direction.

Key *Concepts*

All quantities in the physical and life sciences can be categorized as either scalar or vector quantities. Quantities that can be completely specified by their magnitude (size) are called scalar quantities. Quantities that require specification in both magnitude and direction are called vector quantities

Force vectors and resultant force

When standing upright, there are three forces acting on the human body, body weight W acting at the centre of gravity of the body, the ground reaction force F_L on the left foot and the ground reaction force F_R on the right foot (**Figure 1.22A**). **Figure 1.22B** shows the corresponding vector diagram; the resultant ground reaction force (resultant of F_L and F_R) is equal in magnitude but opposite in direction to W, i.e. the resultant force acting on the body is zero. In **Figure 1.22C** the force F_1 is the resultant ground reaction force (resultant of F_L and F_R) and **Figure 1.22D** shows the corresponding vector diagram. To start walking or running (or move horizontally by any other type of movement such as jumping or hopping) the body must push or pull against something to provide the necessary resultant force to move it in the required direction. In walking and running, forward movement is achieved by pushing obliquely downward and backward against the ground. Provided that the foot does not slip, the leg thrust results in a ground reaction force directed obliquely upward and forward; this is F_2 in **Figure 1.22E**. The resultant force R of W and F_2 moves the body forward while maintaining an upright posture, i.e. the centre of gravity of the body is accelerated in the direction of R. The corresponding vector diagram is shown in **Figure 1.22F**. The vector diagram in **Figure 1.22F** illustrates the vector chain method of determining the resultant of a number of vectors; the component vectors (W and F_2) are linked together

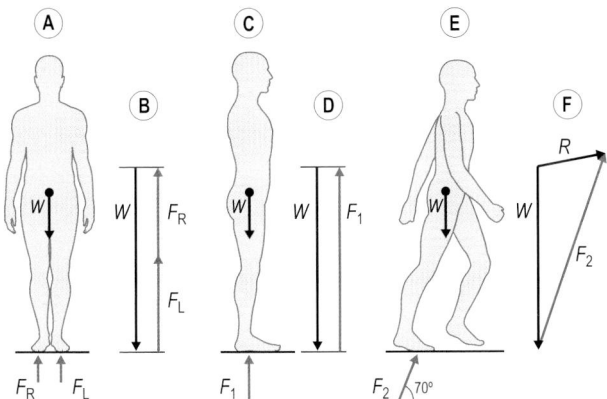

Figure 1.22 (A, B) The forces acting on the body when standing upright (front view) and the corresponding vector diagram. (C, D) The forces acting on the body when standing upright (side view) and the corresponding vector diagram. (E, F) The forces acting on the body just after starting to walk (side view) and the corresponding vector diagram. F_R = force on the right foot, F_L = force on the left foot, W = 70 kgf, F_1 = 70 kgf, F_2 = 80 kgf at 70° to the horizontal, R = 28 kgf at 11° to the horizontal.

in a chain (in any order) and the resultant vector runs from the start of the first component vector to the end of the last component vector. In **Figures 1.22B and D**, all of the forces are vertical forces. Consequently, for clarity, the upward and downward force vectors are presented in parallel (rather than overlapping) in the vector diagrams.

Trigonometry of a right-angled triangle

Whereas the vector chain method of determining the resultant of a number of component vectors is very useful, it is often more practical to use trigonometry, especially when there are a large number of vectors. Trigonometry is the branch of mathematics that deals with the relationships between the lengths of the sides and the sizes of the angles in a triangle. **Figure 1.23** shows a right-angled triangle in which one angle (between sides a and b) is 90°. The angles between sides a and c, and sides b and c, are α and θ, respectively.

In a right-angled triangle, the two angles less than 90° (α and θ in **Figure 1.23**), can be specified by the ratio between the lengths of any two sides of the triangle. The three most common ratios are sine, cosine and tangent, and they are defined in relation to the particular angle under consideration. For example, in relation to angle θ in **Figure 1.23**, side a is

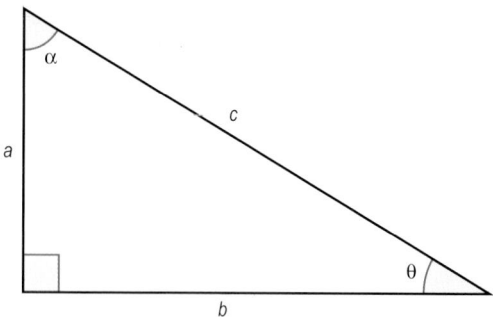

Figure 1.23 Definition of sine, cosine and tangent in a right-angled triangle:

$$\text{sine } \theta = \frac{a}{c} \qquad \text{sine } \alpha = \frac{b}{c}$$
$$\text{cosine } \theta = \frac{b}{c} \qquad \text{cosine } \alpha = \frac{a}{c}$$
$$\text{tangent } \theta = \frac{a}{b} \qquad \text{tangent } \alpha = \frac{b}{a}$$

referred to as the opposite side, side b is referred to as the adjacent side and side c is the hypotenuse, the side of the triangle opposite the right angle.

The sine of θ is defined as the ratio of the opposite side to the hypotenuse, i.e.

$$\text{sine } \theta = \frac{\text{opposite side}}{\text{hypotenuse}} = \frac{a}{c}$$

The cosine of θ is defined as the ratio of the adjacent side to the hypotenuse, i.e.

$$\text{cosine } \theta = \frac{\text{adjacent side}}{\text{hypotenuse}} = \frac{b}{c}$$

The tangent of θ is defined as the ratio of the opposite side to the adjacent side, i.e.

$$\text{tangent } \theta = \frac{\text{opposite side}}{\text{adjacent side}} = \frac{a}{b}$$

Most electronic calculators provide a range of trigonometric ratios including sine (sin), cosine (cos) and tangent (tan). Alternatively, tables of sine, cosine and tangent (for angles between 0° and 90°) can be obtained in publications such as Castle (1969). The lengths of sides and sizes of angles in right-angled triangles can be calculated using sine, cosine and tangent functions provided that two sides or one side and one other angle are known. With reference to **Figure 1.23**, for example, if $c = 10$ cm and $\theta = 30°$, the lengths of sides a and b and the size of angle α can be determined as follows.

1. Calculation of the length of side a

$$\frac{a}{c} = \sin \theta$$
$$a = c.\sin \theta \text{ (i.e. c multiplied by sin } \theta)$$
$$a = c.\sin 30°$$

From sine tables, $\sin 30° = 0.5$ (i.e., the ratio of the length of side a to the length of side c is 0.5). Since $c = 10$ cm and $\sin 30° = 0.5$, it follows that

$$a = 10 \text{ cm} \times 0.5$$
$$a = 5 \text{ cm}$$

2. Calculation of the length of side b

$$\frac{b}{c} = \cos \theta$$
$$b = c.\cos \theta \text{ (i.e. c multiplied by cos } \theta)$$
$$b = c.\cos 30°$$

From cosine tables, $\cos 30° = 0.866$ (i.e., the ratio of the length of side b to the length of side c is 0.866). Since $c = 10$ cm and $\cos 30° = 0.866$, it follows that

$$b = 10 \text{ cm} \times 0.866$$
$$b = 8.66 \text{ cm}$$

3. Calculation of angle α

Angle α can be determined a number of ways:

i. The sum of the three angles in any triangle (with or without a right angle) is 180°. Since the sum of θ and the right angle is 120°, it follows that $\alpha = 180° - 120° = 60°$

ii. The lengths of all three sides of the triangle are known: $a = 5\,cm$, $b = 8.66\,cm$ and $c = 10\,cm$. Consequently, α can be determined by calculating the sine, cosine or tangent of the angle:

$$\sin \alpha = \frac{b}{c} = \frac{8.66\ \text{cm}}{10\ \text{cm}} = 0.866 \quad \alpha = 60°$$

$$\cos \alpha = \frac{a}{c} = \frac{5\ \text{cm}}{10\ \text{cm}} = 0.5 \quad \alpha = 60°$$

$$\tan \alpha = \frac{b}{a} = \frac{8.66\ \text{cm}}{5\ \text{cm}} = 1.732 \quad \alpha = 60°$$

Pythagoras' theorem

Pythagoras, a Greek mathematician (572–497 BC), showed that in a right-angled triangle the square of the hypotenuse is equal to the sum of the squares of the other two sides. Therefore, with respect to **Figure 1.23**

$$c^2 = a^2 + b^2$$
$$c = \sqrt{(a^2 + b^2)}$$

This can be demonstrated with the data from the above example, where $a = 5\,cm$, $b = 8.66\,cm$ and $c = 10\,cm$: $a^2 = 25$, $b^2 = 75$ and $c^2 = 100$.

Resolution of a vector into component vectors

Just as the resultant of any number of component vectors can be determined, any single vector can be replaced by any number of component vectors that have the same effect as the single vector. The process of replacing a vector by two or more component vectors is referred to as the resolution of a vector. In analyzing human movement, displacement, velocity and force vectors are frequently resolved into vertical and horizontal components using trigonometry. The example of the man in **Figure 1.22E** will be used to illustrate how the resolution of forces by trigonometry is used to determine the resultant force acting on the man. There are three steps in the process.

1. Resolve all of the forces into their vertical and horizontal components
   Figure 1.24A shows the forces acting on the man, body weight W and the force F on the right foot. $W = 70\,kgf$ and $F = 80\,kgf$ at 70° to the

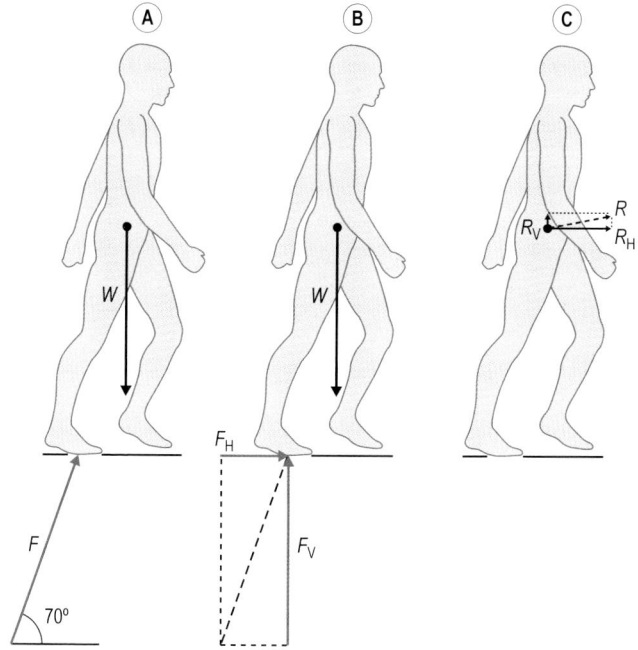

Figure 1.24 The forces acting on a man just after starting to walk. W = body weight = 70 kgf. F = 80 kgf at 70° to the horizontal. R = resultant force = 27.87 kgf at 10.7° to the horizontal.

horizontal. In **Figure 1.24B**, F has been replaced by its vertical F_V and horizontal F_H components. From **Figure 1.24B**

$$F_V/F = \sin 70°$$

$$F_V = F.\sin 70° = 80 \text{ kgf} \times 0.9397 = 75.17 \text{ kgf}$$

$$F_H/F = \cos 70°$$

$$F_H = F.\cos 70° = 80 \text{ kgf} \times 0.3420 = 27.36 \text{ kgf}$$

2. Calculate the vertical component R_V and horizontal component R_H of the resultant force R acting on the man

Using the convention that positive forces act upward and to the right and negative forces act downward and to the left, it follows from Figure 1.24B that

$$R_V = F_V - W$$
$$R_V = 75.17 \text{ kgf} - 70 \text{ kgf} = 5.17 \text{ kgf}$$
$$R_H = F_H = 27.36 \text{ kgf}$$

3. Calculate R and the angle θ of R with respect to the horizontal R, R_V and R_H are shown in Figure 1.24C. From Pythagoras' theorem

$$R^2 = R_V^2 + R_H^2$$
$$R^2 = 26.73 + 748.57 = 775.30$$
$$R = 27.87 \text{ kgf}$$
$$\tan \theta = \frac{R_V}{R_H} = 0.1889$$
$$\theta = 10.7°$$

As expected, R and θ are the same (allowing for error in drawing the vector chain) as in the vector chain solution in Figure 1.22E.

Key *Concepts*

Any single vector can be replaced by any number of component vectors that have the same effect as the single vector. The process of replacing a vector by two or more component vectors is referred to as the resolution of a vector

Moment of a force

Consider a rectangular block of wood resting on a table as shown in Figure 1.25A. The centre of gravity of the block of wood is located at its geometric centre and the line of action of its weight intersects the base of support ABCD. If the block is tilted over on one of the edges of the base of support, such as the edge BC, as in Figure 1.25B, the weight of

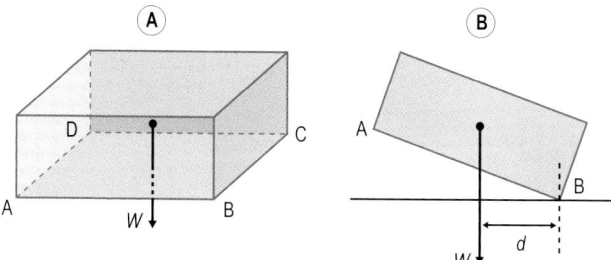

Figure 1.25 The turning moment of a force. (A) Block of wood at rest on base of support ABCD. (B) Turning moment $W.d$ of the weight of the block W tending to restore the block to its original resting position after being tilted over on edge BC. d = moment arm of W about edge BC.

the block W will tend to rotate the block about the supporting edge back to its original resting position in Figure 1.25A. The tendency to restore the block to its original position is the result of the moment (or turning moment) of W about the axis of rotation BC. The magnitude of the moment of W about the axis BC is equal to the product of W and the perpendicular distance d between the axis BC and the line of action of W (Figure 1.25B), i.e.

moment of W about axis AB = $W.d$ (W multiplied by d)

If $W = 2\,\text{kgf}$ and $d = 0.1\,\text{m}$, then

moment of W about axis AB = $2\,\text{kgf} \times 0.1\,\text{m} = 0.2\,\text{kgf.m}$

As $2\,\text{kgf} = 19.62\,\text{N}$, then

moment of W about axis AB = $19.62\,\text{N} \times 0.1\,\text{m} = 1.962\,\text{N.m}$

The N.m (newton metre) is the unit of moment of force in the SI system (see Table 1.1).

In general, when a force F acting on an object rotates or tends to rotate the object about some specified axis, the moment of F is defined as the product of F and the perpendicular distance d between the axis of rotation and the line of action of F, i.e. moment of $F = F.d$. The axis of rotation is often referred to as the fulcrum and the perpendicular distance between the line of action of the force and the axis of rotation is

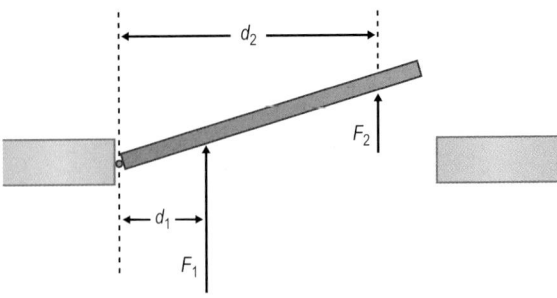

Figure 1.26 Effect of length of moment arm on magnitude of force needed to produce a particular moment of force to open a door. $F_1.d_1 = F_2.d_2$. If $d_2 = 3d_1$, then $F_1 = 3F_2$.

usually referred to as the moment arm of the force. The moment of a force is sometimes referred to as torque. For a given moment of force, the greater the force, the smaller the moment arm of the force and vice versa. For example, in trying to push open a heavy door, much less force will be required if the force is applied to the side of the door furthest away from the hinges, i.e. a large moment arm, than if the force is applied to the door close to the hinges, i.e. a small moment arm (Figure 1.26).

Key *Concepts*

The moment of a force is the product of the magnitude of the force and the perpendicular distance between the line of action of the force and the axis of rotation

Clockwise and anticlockwise moments

When an object is acted upon by two or more forces which tend to rotate the object, the actual amount and speed of rotation that occurs will depend upon the resultant moment acting on the object, i.e. the resultant of all the individual moments.

Figure 1.27 shows the posterior and lateral aspects of the right foot of a runner at foot-strike. In this example, the runner is a rear foot striker and contact with the ground is made with the outside of the heel. In this situation the ground reaction force will tend to be directed upward, laterally and posteriorly. Consequently, when viewed from the rear, the ground reaction force will exert an anticlockwise moment on the heel about an anteroposterior axis through the ankle that tends to evert the heel,

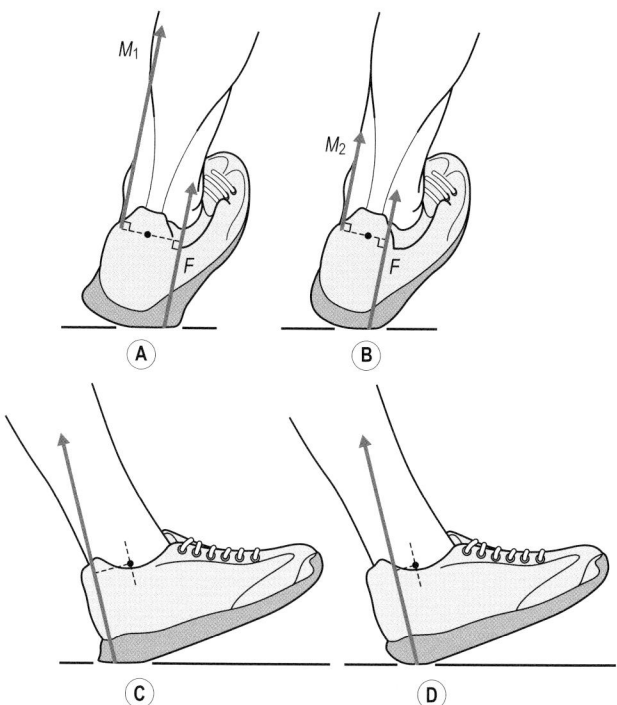

Figure 1.27 Effect of flared heel on the moment arm of the ground reaction force about the ankle joint at heel-strike in running. (A, C) Flared heel, (B, D) non-flared heel. $F =$ ground reaction force, $M_1 =$ force exerted by inversion muscles in flared shoe, $M_2 =$ force exerted by the inversion muscles in the non-flared shoe. The moment arms of M_1 and M_2 are the same.

i.e. rotate the heel in an anticlockwise direction with respect to **Figure 1.27A**. Normally, this tendency to evert the heel is counteracted by a clockwise moment exerted by the muscles that invert the heel, i.e. that rotate the heel in a clockwise direction with respect to **Figure 1.27A**. Consequently, the movement of the heel (clockwise or anticlockwise) will be determined by the resultant moment acting on it, i.e. the resultant of the anticlockwise moment exerted by the ground reaction force and the clockwise moment exerted by the inverter muscles. If these moments were equal in magnitude, then the heel would be held in the same orientation with respect to the ankle, i.e. it would not rotate clockwise or anticlockwise. Following heel-strike, the heel normally everts, i.e. there is a resultant eversion moment acting on it during the period between heel-strike and foot-flat. The magnitude of the resultant eversion moment determines the amount of eversion (angular distance) and angular speed of eversion. Both the amount and speed of eversion must be carefully controlled in order to reduce the risk of injury. Control of eversion depends largely on the capacity of the inversion muscle moment to counteract the ground reaction force moment. The higher the ground reaction force moment, the greater the load on the inversion muscles and vice versa. The magnitude of the moment arm of the ground reaction force and, therefore, the magnitude of the ground reaction force moment will depend upon the shape of the heel of the shoe. Figure 1.27A shows that a lateral flared heel tends to increase the moment arm of the ground reaction force compared to a non-flared shoe (Figure 1.27B). If the moment arm of the ground reaction force in Figure 1.27A is double that in Figure 1.27B, then the eversion moment exerted by the ground reaction force in Figure 1.27A will be double that in Figure 1.27B. As the moment arm of the inversion muscles is unlikely to change in the two different types of shoe, it follows that the load on the inversion muscles in the flared shoe will be double that in the non-flared shoe. Similarly, the load on the ankle dorsiflexors about a mediolateral axis through the ankle in controlling plantar flexion between heel-strike and foot-flat is likely to be much greater in a flared shoe than in a non-flared shoe (Figure 1.27C and D).

Figure 1.28A shows two boys A and B sitting on a seesaw S. A seesaw is normally constructed so that its centre of gravity coincides with the fulcrum, i.e. in any position, the line of action of its weight will pass through the fulcrum and, therefore, the weight of the seesaw will not exert a turning moment on the seesaw since the moment arm of its weight about the fulcrum will be zero. Consequently, in Figure 1.28A the only moments tending to rotate the seesaw will be those exerted by the weights of the two boys. The weight of A will exert an anticlockwise moment $W_A \times d_A$ and the weight of B will exert a clockwise moment $W_B \times d_B$. When $W_A \times d_A$ is greater than $W_B \times d_B$, there will be a resultant anticlockwise moment acting on the seesaw such that B will be lifted as A descends. When $W_A \times d_A$ is equal to $W_B \times d_B$, i.e. when the clockwise moment is equal to the anticlockwise moment, the resultant moment acting on the seesaw will be zero and the seesaw will not rotate in either direction, but remain perfectly still in a balanced position. Consequently, if the weight of one of the boys was known, the weight of the other boy could be found by balancing the seesaw with one boy on each side of the fulcrum with both boys off the floor and then equating the clockwise and anticlockwise moments. For example, if $W_A = 40\,kgf$ and in the balanced

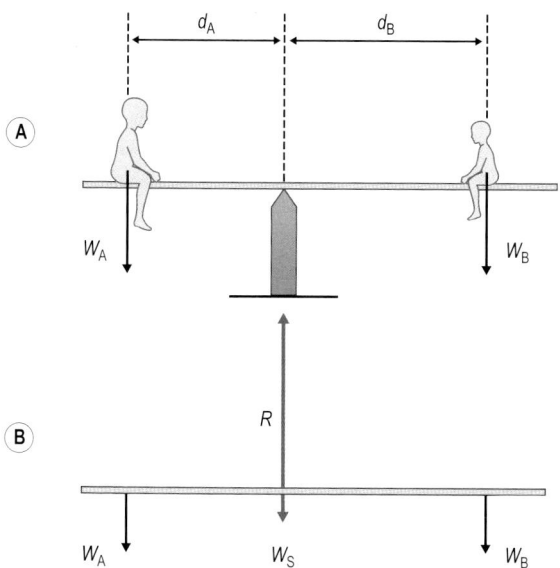

Figure 1.28 (A) Two boys sitting on a seesaw. (B) Free body diagram of the seesaw. W_A = weight of boy A; W_B = weight of boy B; d_A = moment arm of W_A; d_B = moment arm of W_B; W_S = weight of seesaw; R = force exerted by the seesaw support on the seesaw = $W_A + W_B + W_S$.

position $d_A = 1.5\,$m and $d_B = 2.0\,$m, then by equating moments about the fulcrum

anticlockwise moments (ACM) = clockwise moments (CM)

$$40 \text{ kgf} \times 1.5 \text{ m} = W_B \times 2.0 \text{ m}$$

$$W_B = \frac{40 \text{ kgf} \times 1.5 \text{ m}}{2.0 \text{ m}} = 30 \text{ kgf}$$

Key *Concepts*

When an object is acted upon by two or more forces which tend to rotate the object, the amount and speed of rotation that occurs will depend upon the resultant moment

Equilibrium

In the situation illustrated in **Figure 1.28A**, the moments of the weights of the two boys acting about the fulcrum of the seesaw are equal and opposite such that the resultant moment acting on the seesaw is zero and the seesaw is at rest. As the seesaw is at rest, the resultant force on the seesaw is also zero. Consequently, the resultant downward force on the seesaw, i.e. the weights of the seesaw and the two boys, must be counteracted by one or more forces whose resultant is equal and opposite to the weights of the seesaw and the two boys. In this case, the counteracting force is a single force R exerted by the seesaw support through the fulcrum. **Figure 1.28B** shows a free body diagram of the seesaw (a diagram showing all of the external forces acting on the seesaw) in the situation illustrated in **Figure 1.28A**. As $W_A = 40$ kgf, $W_B = 30$ kgf and $W_S = 20$ kgf, then $R = W_A + W_B + W_S = 90$ kgf.

With regard to linear motion, an object is in equilibrium when the resultant force acting on the object is zero. With regard to angular motion, an object is in equilibrium when the resultant moment acting on the object is zero. These two equilibrium conditions are the basis for calculating forces acting on different parts of a multi-segment system, such as muscle forces and joint reaction forces in the human body.

Achilles tendon force and ankle joint reaction force in upright standing

Figure 1.29A shows a man standing upright with the line of action of his weight slightly in front of his ankle joints. Consequently, body weight will exert an anticlockwise moment about the axis of plantar flexion-dorsiflexion (A) tending to rotate his body forward. In this posture, stability is maintained largely by isometric contraction of the ankle joint plantar flexors, as shown in the simple two-segment model in **Figure 1.29B** where W = body weight, R = the ground reaction force and F = the force in the Achilles tendons (both legs). If it is assumed that each leg supports half of body weight, **Figure 1.29C** shows a free body diagram of one foot where R_1 = ground reaction force, T = force in the Achilles tendon acting an angle of 85° to the horizontal and J = ankle joint reaction force. As the weight of the foot ($\approx 1.45\%$ of total body weight) is very small relative to T and J, and has a negligible effect on T and J, it is not included in the free body diagram of the foot in **Figure 1.29C**. J acts through A and, as such, exerts no moment about A. As the foot is in equilibrium, the

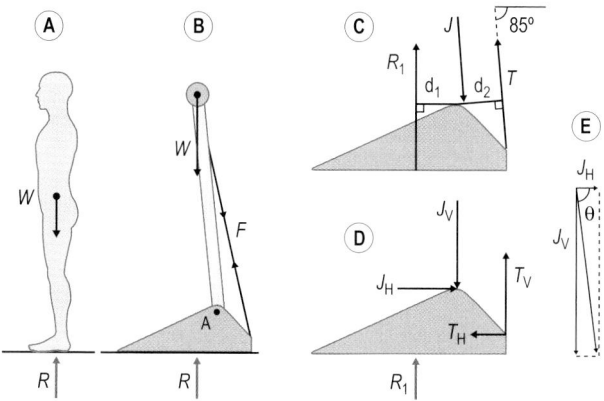

Figure 1.29 Achilles tendon force and ankle joint reaction force when standing upright.

resultant moment on the foot is zero. Consequently, taking moments about A, it follows that

$$R_1.d_1 = T.d_2$$

Eq. 1.2

where d_1 = moment arm of R_1 about A and d_2 = moment arm of T about A.

If W = 70 kgf, then R_1 = 35 kgf. When d_1 = 5 cm, d_2 will also be about 5 cm. Consequently, from equation 1.2

$$T = \frac{R_1.d_1}{d_2} = \frac{35 \text{ kgf} \times 5 \text{ cm}}{5 \text{ cm}} = 35 \text{ kgf}$$

Figure 1.29D shows a free body diagram of the foot in which J and T have been replaced by their horizontal (J_H, T_H) and vertical (J_V, T_V) components. As T acts at 85° to the horizontal then

$$T_H/T = \cos 85°$$

$$T_H = T.\cos 85° = 35 \text{ kgf} \times 0.0871 = 3.05 \text{ kgf}$$

$$T_V/T = \sin 85°$$

$$T_V = T.\sin 85° = 35 \text{ kgf} \times 0.9962 = 34.87 \text{ kgf}$$

As the foot is in equilibrium, the resultant force on the foot is zero. Consequently, using the convention that positive forces act upward and to the right and negative forces act downward and to the left, it follows that

$$R_1 + T_V - J_V = 0$$
$$J_V = R_1 + T_V = 35 \text{ kgf} + 34.87 \text{ kgf} = 69.87 \text{ kgf}$$
$$J_H - T_H = 0$$
$$J_H = T_H = 3.05 \text{ kgf}$$

From Pythagoras' theorem

$$J^2 = J_V^2 + J_H^2$$
$$J^2 = 4881.82 + 9.30 = 4891.12$$
$$J = 69.93 \text{ kgf}$$

From **Figure 1.28E**

$$\tan \theta = \frac{J_V}{J_H} = 22.908$$
$$\theta = 87.5°$$

Consequently, when standing upright with the line of action of body weight about 5 cm in front of the ankle joint, the Achilles tendon force (tension) and ankle joint reaction force (compression) in each foot will be approximately half body weight and body weight, respectively. Leaning further forward will increase d_1, but d_2 will hardly change. Therefore, from equation 1.2, it is clear that leaning further forward will increase the Achilles tendon force which, in turn, will increase the ankle joint reaction force. When leaning forward with the line of action of body weight acting through the balls of the feet, the Achilles tendon force and ankle joint reaction force in each foot will be approximately $1.3W$ and $1.8W$, respectively.

Levers

Whereas weight forces are always vertical, other external forces acting on a body are likely to be oblique. For example, consider using a claw

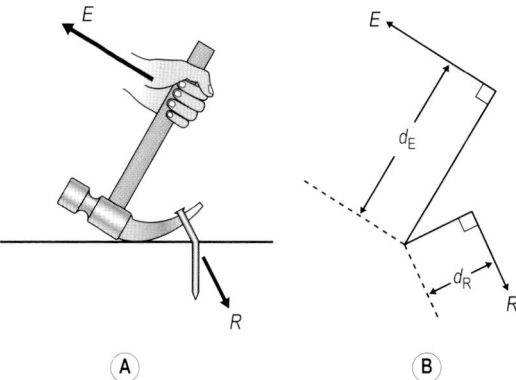

Figure 1.30 (A) Pulling a nail out of a piece of wood using a claw hammer. (B) The corresponding force-moment arm diagram. E = force exerted on the handle of the hammer; d_E = moment arm of E; R = resistance force exerted by the nail; d_R = moment arm of R.

hammer to pull out a nail from a piece of wood as shown in **Figure 1.30A**. In this example, the line of contact between the claw of the hammer and the surface of the wood constitutes the fulcrum. E is the force exerted on the handle of the hammer and R is the resistance of the nail. **Figure 1.30B** shows a force-moment arm diagram, i.e. the forces E and R and their moment arms d_E and d_R are shown in relation to the fulcrum in order to more clearly illustrate the turning effects of the forces. The nail will be pulled out if the anticlockwise moment exerted by E is greater than the clockwise moment exerted by R, i.e. if $E.d_E > R.d_R$.

In this situation the hammer is being used as a lever, a rigid or quasi-rigid object that can be made to rotate about a fulcrum in order to exert a force on another object. As in the example of the hammer, a lever encounters a resistance force R in response to an effort force E. The simplest form of lever, which is actually the simplest form of machine, i.e. a powered mechanism designed to apply force (Dempster 1965), is exemplified by a crowbar as shown in **Figure 1.31**. In this case, the power is supplied by the person using the crowbar. The greater the moment arm of E (d_E), i.e. the greater the leverage of the crowbar, the smaller will be the effort required to overcome the moment of the resistance force.

The bones of the skeleton are essentially levers and each joint constitutes a fulcrum. The muscles pull on the bones to control the movement of the joints. The resistance to movement exerted by a body segment is in the form of the segment's weight and any other external loads attached to the segment.

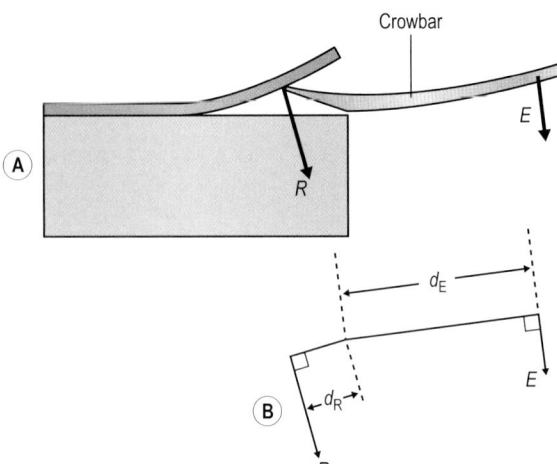

Figure 1.31 (A) Use of a crowbar to open a box. (B) The corresponding force-moment arm diagram. E = force exerted on the crowbar by the person using it; d_E = moment arm of E; R = resistance force exerted on the crowbar by the lid of the box; d_R = moment arm of R.

Key Concepts

A lever is an object that can be made to rotate about a fulcrum in order to exert a force on another object

Mechanical advantage

The mechanical advantage (MA) of a lever is a measure of its efficiency in terms of the amount of effort needed to overcome a particular resistance, i.e.

$$\text{MA} = \frac{\text{magnitude of resistance}}{\text{magnitude of effort}} = \frac{R}{E}$$

$$= \frac{\text{length of moment arm of } E}{\text{length of moment arm of } R} = \frac{d_E}{d_R}$$

The greater the mechanical advantage of a lever, the smaller the effort needed to overcome a particular resistance.

Load on the musculoskeletal system

The open-chain arrangement of the bones (four peripheral chains – the arms and legs – attached onto a central chain – the vertebral column) and skeletal muscles (closely aligned to the bones) maximizes the range of possible body postures. However, the price that the body pays for this movement capability is low mechanical advantage. Most of the skeletal muscles are inserted close to the joints that they control and, as such, have shorter moment arms than the resistance forces they counteract, i.e. the skeletal muscles usually have to exert much larger forces than the weights of the body segments they control. Furthermore, as illustrated in **Figure 1.29**, joint reaction forces are determined by muscle forces; the larger the muscle forces, the larger the joint reaction forces. For example, in walking, the peak hip, knee and ankle joint reaction forces in adults are normally in the range of 5 to 6, 3 to 8, and 3 to 5 times body weight, respectively (Nigg 1985). The more dynamic the activity, the greater the muscle forces and, therefore, the greater the joint reaction forces. For example, in fast running (800 m pace), the peak knee and ankle joint reaction forces in an adult are likely to be in the region of 20 and 8 times body weight, respectively (Nigg 1985).

In any body position other than the relaxed recumbent position, the components of the musculoskeletal system (muscles, bones, joints) are likely to be subjected to considerable load. In response to the load the musculoskeletal components experience strain; the greater the force, the greater the strain. Under normal circumstances the musculoskeletal components adapt their size, shape and structure to the time-averaged forces exerted on them in order to more readily withstand the strain. However, excessive strain will result in injury and/or degeneration. Consequently, there is an intimate relationship between the structure and function of the musculoskeletal system (Watkins 1999).

Key Concepts

The open-chain arrangement of the skeleton maximizes the range of possible body postures, but only at the expense of low mechanical advantage; the muscles, bones and joints are subjected to very high forces in virtually all postures other than lying down

Review questions

1. Define the following terms: force, contact force, attraction force, resultant force, mechanics, biomechanics, mass, inertia, volume, density, stiffness, strength, kinematics, kinetics.
2. Describe the two ways in which forces tend to affect bodies.
3. Describe the four main sub-disciplines of mechanics.
4. Describe the two fundamental forms of motion.
5. List the base units for length, mass and time in the International System of Units.
6. With regard to musculoskeletal system function, describe the relationship between external and internal forces.
7. Briefly describe the three broad categories of movement brought about by the musculoskeletal system.
8. With reference to recumbent posture and standing posture, explain the difference between direct and indirect transmission of the weight of body segments to the support surface.
9. Describe how direct and indirect transmission of the weight of body segments to the support surface likely affects the degree of activity in the muscles.
10. Describe the main advantage and disadvantage of the open-chain arrangement of the skeleton.

References

Bagga H, Burkhardt D, Sambrook P, March L (2006) Longterm effects of intraarticular hyaluronan on synovial fluid in osteoarthritis of the knee. Journal of Rheumatology 33:946–50.

Castle F (1969) Five-figure logarithmic and other tables. Macmillan, London.

Divine PG, Zazulak BT, Hewett TE (2007) Viscosupplementation for knee osteoarthritis: a systematic review. Clinical Orthopaedics and Related Research 455:113–22.

Dempster WT (1965) Mechanisms of shoulder movement. Archives of Physical Medicine and Rehabilitation 46:49–70.

eOrthopod (2008) Viscosupplementation for osteoarthritis of the knee. Available online at: http://eorthopod.com/public/patient_education/6515/viscosupplementation for the knee.html

Elert G, ed (2000) The physics factbook. Available online at: http://hypertextbook.com/facts/2000/KatherineMalfucci.shtml

Hennig EM, Staats A, Rosenbaum D (1994) Plantar pressure distribution patterns of young children in comparison to adults. Foot and Ankle 15(10):35–40.

Nigg BM (1985) Biomechanics, load analysis, and sports injuries in the lower extremities. Sports Medicine 2:367–79.

Rowett R (2004) How many? A dictionary of units of measurement. Available online at: http://www.unc.edu/~rowlett/units/sipm.html 14 June 2004

Roberts TDM (1995) Understanding balance: the mechanics of posture and locomotion. Chapman & Hall, London.

Serway RA, Jewett JW (2004) Physics for scientists and engineers. Harcourt College Publishers, Fort Worth, TX.

Watkins J (1999) Structure and function of the musculoskeletal system. Human Kinetics, Champaign, IL.

Watkins J (2007) An introduction to biomechanics of sport and exercise. Churchill Livingstone, Edinburgh.

Tissues, organs and systems

All living organisms are made up of cells. The human body is made up of billions of cells that are organized into complex groups that carry out specific functions. These groups include the cardiovascular system, which transports blood around the body, and the musculoskeletal system, which enables us to move. The purpose of this chapter is to describe the different types of cell in the body and the organization of the cells within functional groups.

Unicellular and multicellular organisms

Cells are the building blocks of life, i.e. the fundamental structural and functional units of all living organisms. The lowest forms of life consist entirely of single cells; these organisms are referred to as unicellular organisms. Higher forms of life, like the human body, consist of many cells; these organisms are referred to as multicellular organisms.

There are two general categories of cell, prokaryotes and eukaryotes (Alberts 2002). Prokaryotes, which include bacteria, were the first types of unicellular organism to evolve. A prokaryotic cell consists of an outer cell membrane (comprised of proteins and lipids) that encloses a semitransparent fluid called cytosol (a complex solution of proteins, salts and sugars). The cell membrane, usually referred to as the plasma membrane or plasmalemma, separates and protects the cell from its surrounding environment and allows interchange of substances between the cytosol and the surrounding environment via a system of channels and pumps.

Eukaryotes are similar to prokaryotes in that they have a cell membrane that encloses cytosol. However, there are significant differences in structure and function. Whereas the cytosol of a prokaryotic cell is largely featureless, the cytosol of a eukaryotic cell surrounds a complex structure of membrane-bound compartments and other structures called organelles (**Figure 2.1**). The most distinct membrane-bound compartment is the nucleus. The nucleus contains the cell's genetic material, which organises and controls, via the organelles, the life processes of the cell. The life processes include growth and development, respiration, circulation, digestion, excretion, reproduction and movement. The life processes in prokaryotes are carried out by indistinct structures in the cell membrane. Whereas prokaryotes only produce unicellular organisms, eukaryotes produce unicellular and multicellular organisms. All mammals, birds and fish are multicellular organisms. The number of cells in multicellular organisms varies considerably. For example, the nematode worm

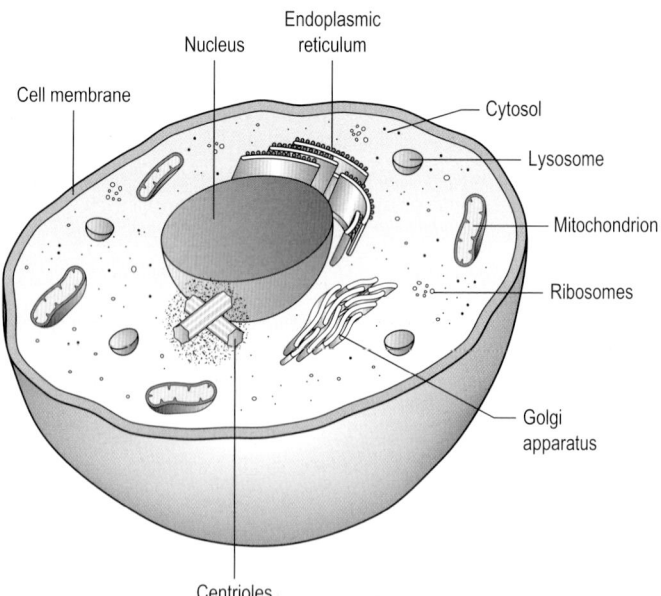

Figure 2.1 Section through a generalized eukaryotic cell.

C.elegans is 1mm in length and consists of 959 cells (Kenyon 1988). Most multicellular organisms consist of millions of cells; the human body consists of approximately 10^{14} (one hundred million million) cells (Alberts 2002).

Key Concepts

The fundamental structural and functional unit of living organisms is the cell. Unicellular organisms consist entirely of a single cell. Multicellular organisms consists of many cells

Like cytosol, the cell membrane consists of a viscous fluid, but it is usually much more viscous than the cytosol. The viscous nature of the cell membrane and cytosol enables a cell to change its shape without losing its integrity. During normal functioning, many cells regularly change shape and the amount of change depends upon the type of cell. For example, in muscle cells the ability to change shape is highly specialized and is an essential feature of normal function. Muscle cells are long and thin and may shorten by up to 40% during contraction. In contrast, bone cells, for example, occupy tiny spaces within a fairly hard bony matrix such that change in shape of bone cells only occurs very gradually as a result of bone growth.

The structure of a cell determines its function. Whereas muscle cells and bone cells are both eukaryotic, they have different functions and, as such, are different in structure. The human body is made up of four fundamental types of eukaryotic cell called tissues.

Cellular organization in multicellular organisms

All living organisms carry out all of the essential life processes. In unicellular organisms, the life processes are, in biological terms, relatively simple. However, in multicellular organisms, the cells are organized into complex functional groups that carry out the various life processes for the body as a whole (Watkins 1999). There are three levels of cellular organization in each functional group: tissues, organs and systems.

Tissues

In multicellular organisms all of the cells originate from a single cell formed by the fertilization of a female ovum by a male sperm. This cell undergoes rapid cell division to form a large number of similar cells. Soon after this, the cells differentiate in size, shape and structure in order to carry out different functions within the developing organism. This process of cellular differentiation results in the formation of four types of cell called tissues: epithelia, nerve, muscle and connective. A tissue is a group of cells having the same specialized structure, enabling them to perform a particular function in the body (Freeman and Bracegirdle 1967). The word tissue is also used in a general sense such as, for example, the description of skin, muscles, tendons, ligaments and fat as soft tissues (Watkins 1999).

Epithelial tissue

There are two types of epithelial tissue: covering and glandular. Covering epithelia form the surface layer(s) of cells of all the internal and external free surfaces of the body except the surfaces inside synovial joints. For example, the surfaces of the skin and the lining of the alimentary canal, heart chambers and blood vessels are all examples of covering epithelia (Watkins 1999).

All cells secrete fluid to a greater or lesser extent, but glandular epithelial cells are specialized for this purpose and as such form the two types of gland found in the body: exocrine and endocrine. Many of the exocrine glands (glands with ducts that discharge secretions onto a surface), such as the gastric glands in the lining of the stomach, secrete fluids containing enzymes necessary for the digestion of food. Endocrine glands (ductless glands that discharge secretions directly into the blood stream), such as the pituitary at the base of the brain and the adrenals at the upper end of each kidney, secrete hormones that, in association with the nervous system, regulate and coordinate the various body functions (Watkins 1999).

Nerve tissue

Nerve cells (neurones) are specialized to conduct electrochemical impulses throughout the body to coordinate the various body functions. The structure and functions of nerve tissue are covered in detail in Chapter 6 (The neuromuscular system).

Muscle tissue

Muscle cells are specialized to contract, i.e. create a pulling force to stabilize or move parts of the body. There are three types of muscle cell: skeletal, visceral and cardiac.

Skeletal muscle is attached to the skeletal system (skeleton and joint support structures) and, under the control of the nervous system, controls movement in the joints between the bones of the skeleton. Skeletal muscle is also referred to as voluntary muscle because it is normally under the conscious control of the individual. The structure and functions of skeletal muscle are covered in detail in Chapter 6 (The neuromuscular system).

Visceral muscle, also referred to as involuntary muscle because it is not normally under conscious control, is found in parts of the body that experience involuntary movement such as the walls of the larger arteries and alimentary canal. For example, contraction of visceral muscle in the walls of an artery reduces the diameter of the artery and results in a local increase in blood pressure that pushes blood around the cardiovascular system. Similarly, contraction of visceral muscle in the walls of the alimentary canal pushes digested food along the canal.

Cardiac muscle is found only in the heart. It has characteristics of both skeletal and visceral muscle, but differs from them in that it contracts rhythmically throughout life even though the frequency of contractions (heart rate) varies.

Connective tissue

As its name suggests, one of the main functions of connective tissue is to support and bind other tissues together. The bones of the skeleton and the fibrous structures that hold the bones together at joints are all forms of connective tissue. The structure and functions of connective tissue are covered in detail in Chapter 4 (Connective tissues).

Key *Concepts*

> Cellular differentiation results in the formation of four types of cell called tissues: epithelia, nerve, muscle and connective

Organs and systems

An organ is a combination of different tissues designed to carry out a specific bodily function. For example, the heart is designed to pump blood around the body. The structure of the heart consists of cardiac muscle cells, connective tissue which binds the muscle cells together, epithelial tissue which lines the chambers of the heart, and nerve tissue which innervates the cardiac muscle cells (Watkins 1999). Other examples of organs are the brain, the eyes, the lungs, the stomach, the spleen, the kidneys, the liver and each skeletal muscle complete with its tendons that attach the muscle to the skeletal system.

A system is a combination of different organs working together to carry out a particular bodily function. For example, the function of the cardiovascular system, which consists of the heart and blood vessels, is to transport blood around the body. There are basically 11 separate systems in the human body (Tortora and Anagnostakos 1984):

Integumentary system: the external covering of the body, i.e. the skin and associated structures such as finger and toe nails.

Skeletal system: the bones of the skeleton and the structures that form the joints between the bones.

Muscular system: the skeletal muscles.

Nervous system: the nerves, organized into central (brain and spinal cord) and peripheral (spinal nerves) components.

Endocrine system: the glands that secrete hormones which regulate and coordinate the various body functions in association with the nervous system.

Cardiovascular system: the heart and blood vessels.

Lymphatic system: the system of vessels and ducts that drains extracellular fluid and returns it to the blood.

Respiratory system: the lungs and associated passageways.

Digestive system: the alimentary canal and associated structures that break down food and eliminate solid waste.

Urinary system: the kidneys, bladder and associated structures that eliminate nitrogenous waste as urine.

Reproductive system: the ovaries and associated structures (female) and testes and associated structures (male) that enable the body to produce offspring.

These systems are responsible for carrying out the body's life processes. Whereas all of the life processes involve a certain degree of integration between different systems, some processes involve closer integration than others. For example, the transport of oxygen from the air to all the cells of the body and the transport of carbon dioxide in the opposite direction is carried out by close integration of the nervous, respiratory, and cardiovascular systems. Similarly, movement of the body is brought about by close integration of the nervous, muscular, and skeletal systems. Consequently, the systems are often referred to in combination as, for example, cardiorespiratory system, musculoskeletal system and neuromuscular system. **Figure 2.2** summarizes cellular differentiation and organization in multicellular organisms (Watkins 1999).

Key Concepts

Cellular organization results in the formation of organs (combinations of different tissues) and systems (combinations of different organs). The systems are responsible for carrying out the body's life processes

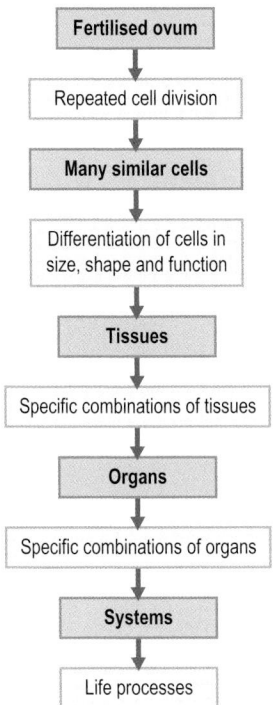

Figure 2.2 Cellular differentiation and organization in multicellular organisms.

The musculoskeletal system

The musculoskeletal system consists of the skeletal system and the muscular system. The skeletal system consists of the skeleton (the bones) and the fibrous structures that form the joints between the bones. Figure 2.3 shows the skeleton and the muscular system. The skeleton gives the body its basic shape and provides a very strong, relatively lightweight supporting framework for all the other systems. The adult skeletal system normally has 206 bones, more than 200 joints and accounts for approximately 12% and 15% of total body weight in females and males, respectively (McArdle et al 1996). The structure and function of the skeleton are covered in detail in Chapter 3 (The skeleton).

As previously described, there are three types of muscle tissue: visceral, cardiac and skeletal. Visceral muscle is usually considered to be part

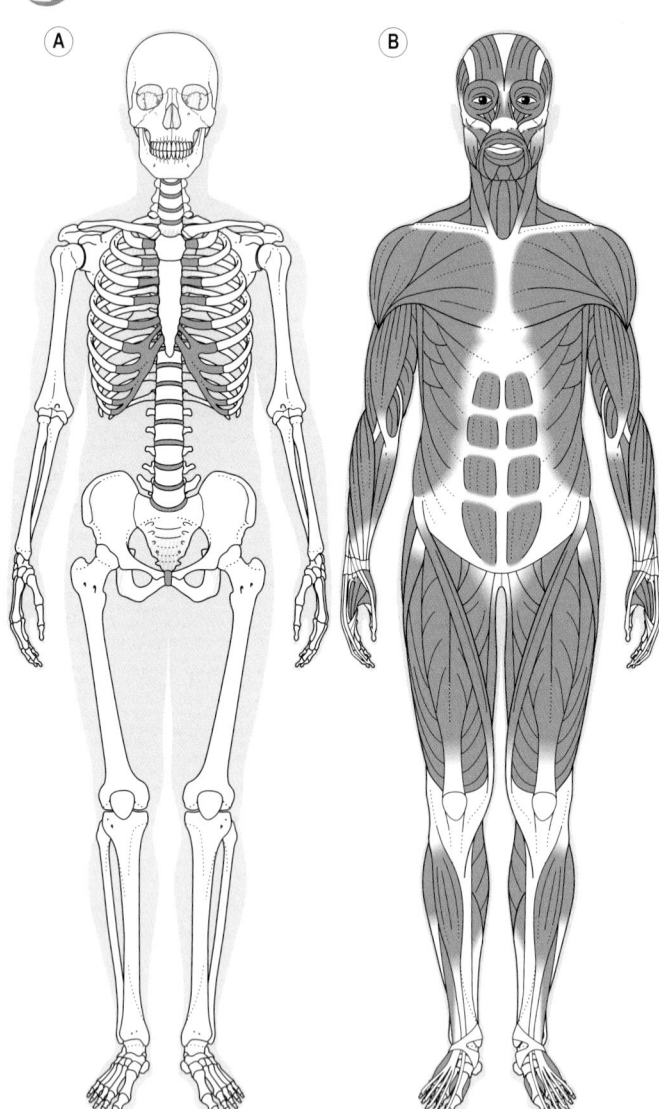

Figure 2.3 (A) The skeleton. (B) The muscular system.

of the digestive (in the walls of the alimentary canal) and cardiovascular (in the walls of arteries) systems and cardiac muscle is part of the cardiovascular system (Tortora and Anagnostakos 1984). Consequently, the muscular system refers only to the skeletal muscles. There are approximately 640 skeletal muscles, which account for an average of approximately 34% and 42% of total body weight in young (18–29 years) healthy untrained females and males, respectively, and an average of approximately 30% and 34% of total body weight in elderly (70–88 years) healthy untrained females and males, respectively (Janssen et al 2000).

Key *Concepts*

The musculoskeletal system consists of the skeletal system (the bones and the fibrous structures that form the joints between the bones) and the muscular system (the skeletal muscles)

The most important property of all types of muscle tissue is contractility, i.e. the ability to create a pulling force and, if necessary, change in length (increase or decrease) while maintaining a pulling force. This property is developed to a high level in skeletal muscle tissue. Each complete skeletal muscle consists of a large number of longitudinal-shaped muscle cells bound together by various layers of connective tissues. The length of the cells varies from a few millimetres (in the muscles that move the eyes) to about 30 cm (sartorius muscle in the anterior thigh). In most skeletal muscles the muscle cells occupy the main belly of the muscle, but the ends of the muscle consist entirely of thickened cords or bands of virtually inextensible connective tissue that anchor the muscle onto the skeleton (**Figure 2.4**). The shape of these connective tissue attachments depends on the shape of the muscle and the attachment areas available on the skeleton. In general, there are two basic shapes: a cord or narrow band called a tendon and a broad band called an aponeurosis (see **Figure 2.4**). A skeletal muscle and its tendons and aponeuroses are usually referred to as a musculotendinous unit (Taylor et al 1990).

As a muscle can only pull in one direction, the movement of each joint is controlled by opposing pairs of muscle groups, referred to as antagonistic pairs. The muscles in each group cross over one or more joints. For example, the quadriceps muscle group consists of four muscles: vastus lateralis, vastus intermedius, vastus medialis and rectus femoris (**Figure 2.5**). The rectus femoris originates from the front of the pelvis and the vasti muscles originate from the shaft of the femur. The four muscles join together to form the quadriceps tendon, which covers the anterior

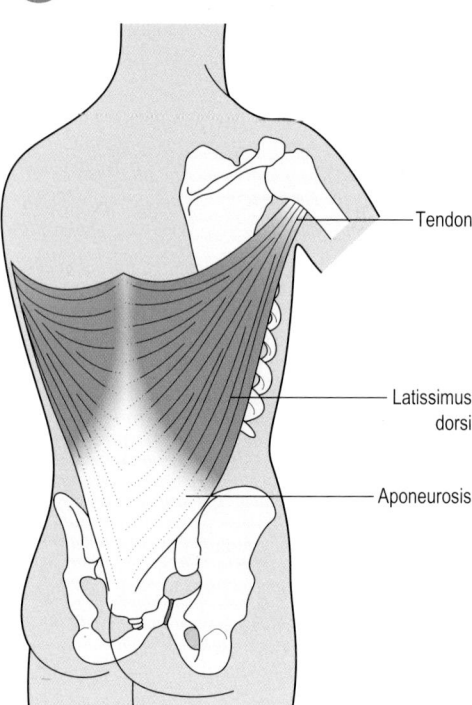

Tendon

Latissimus dorsi

Aponeurosis

Figure 2.4 Tendon and aponeurosis. The latissimus dorsi muscle on each side of the body originates from a broad aponeurosis and inserts onto the humerus via a narrow-band tendon.

aspect of the patella and continues as the patella ligament to attach onto the tibial tuberosity of the tibia. Consequently, all four muscles cross the front of the knee joint and tend to extend the knee when the quadriceps contract. However, the rectus femoris also crosses the front of the hip joint and tends to flex the hip when the quadriceps contract. The rectus femoris is a two-joint muscle and the vasti muscles are one-joint muscles. The hamstrings muscle group is antagonistic to the quadriceps and consists of three muscles: semimembranosus, semitendinosus and biceps femoris. The semimembranosus, semitendinosus and part of the biceps femoris all originate from the ischial tuberosity and insert onto the tibia, i.e. they are two-joint muscles that tend to extend the hip and flex the knee. Part of the biceps femoris originates from the shaft of the femur, i.e. part of the biceps femoris is a one-joint muscle (that tends to flex the knee) and the other part is a two-joint muscle.

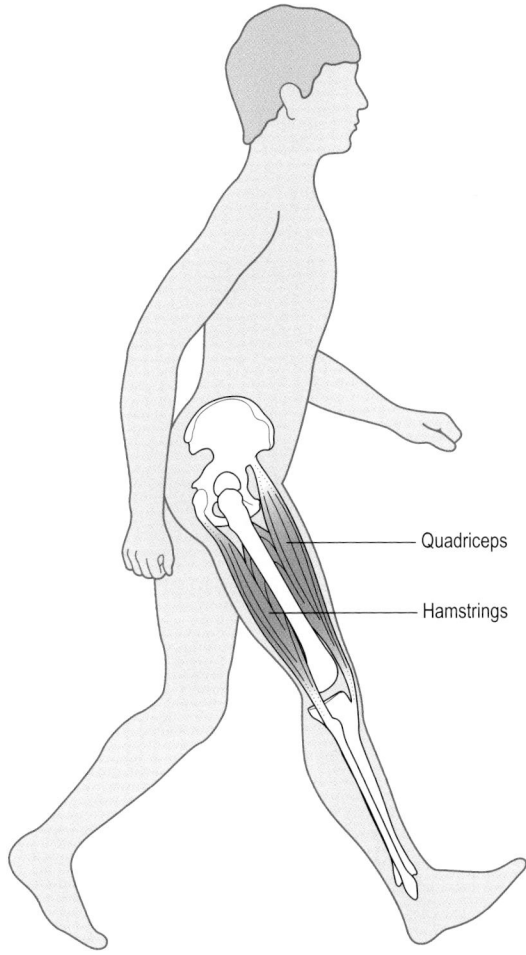

Quadriceps

Hamstrings

Figure 2.5 Antagonistic muscle groups: quadriceps and hamstrings.

By coordinated activity between the groups of muscles that cross a joint, the amount and rate of movement in the joint can be carefully controlled. The number of pairs of muscles required to control a particular joint depends upon the number of directions in which the joint is free to move. For example, the knee joint is basically designed to rotate in one plane about a transverse (side-to-side) axis (knee flexion/extension).

However, other joints such as the hip and shoulder are designed to move in more than one plane (see Chapter 5). Consequently, two or more antagonistic pairs of muscle groups are required to control movement in these joints.

Key Concepts

Each skeletal muscle is attached to the skeleton such that it crosses one or more joints; when a muscle contracts it tends to bring about movement in the joints it crosses. The muscles are arranged in antagonistic pairs that coordinate action to control the amount and rate of movement in each joint

Review questions

1. Differentiate between cellular differentiation and cellular organization in multicellular organisms.
2. Describe the four basic tissues.
3. Describe what is meant by a combined system.
4. Describe what is meant by an antagonistic pair of skeletal muscles.

References

Alberts B, Johnson A, Lewis J, et al (2002) Molecular biology of the cell, 4th ed. Garland Publishing, New York.

Freeman WH, Bracegirdle B (1967) An atlas of histology. Heinemann, London.

Kenyon C (1988) The nematode Caenorhabditis elegans. Science 240:1448–52.

Janssen I, Heymsfield SB, Wang Z, Ross R (2000) Skeletal muscle mass and distribution in 468 men and women aged 18–88 yr. Journal of Applied Physiology 89:81–88.

McArdle WD, Katch FI, Katch VL (1996) Exercise physiology: energy, nutrition, and human performance. Lea & Febiger.

Taylor DC, Dalton JD, Seaber AV, Garrett WE (1990) Viscoelastic properties of muscle-tendon units. American Journal of Sports Medicine 18:300–8.

Tortora GJ, Anagnostakos NP (1984) Principles of anatomy and physiology. Harper & Row, New York.

Watkins J (1999) Structure and function of the musculoskeletal system. Human Kinetics, Champaign, IL.

The skeleton

The skeleton gives the body its basic shape
and provides a strong, protective and supporting framework for all other systems of the
body. Mechanically, the bones are levers oper-
ated by the skeletal muscles. The mechanical
advantage of the levers in the musculoskeletal
system varies considerably; this is reflected in
the wide variety in size and shape of the bones.
The purpose of this chapter is to describe the
bones of the skeleton and, in particular, the features of the bones associated with force transmission and relative motion between bones.

Composition and function of the skeleton

The mature skeleton normally consists of 206
bones. However, variations in this basic number
do occur; for example, some adults have 11
or 13 pairs of ribs, whereas most adults have
12 pairs. The skeleton performs three main
mechanical functions (Watkins 1999):

1. It is a supporting framework for the rest of
   the body.
2. It is a system of levers on which the muscles can pull in order to stabilize and move
   the body.
3. It protects certain organs. For example,
   the skull protects the brain, the vertebral
   column protects the spinal cord and the rib
   cage helps protect the heart and lungs.

Terminology

For descriptive purposes the bones are usually divided into two main groups: the axial skeleton and the appendicular skeleton (axial = axis, appendicular = appendage). The adult axial skeleton consists of 80 bones comprising the skull, vertebral column and ribs. The adult appendicular skeleton consists of 126 bones that make up the upper limbs (arms and hands) and the lower limbs (legs and feet) (Figure 3.1).

In anatomy, the term aspect refers to appearance from a particular viewpoint. For example, the anterior aspect of the skeleton refers to an anterior (frontal) view (Figure 3.1(A). Similarly, lateral aspect (view from the side) (Figure 3.1B), posterior aspect (view from the back), superior aspect (view from above) and inferior aspect (view from below) describe other views.

The bones vary considerably in size and shape. The smallest mature bones are the auditory ossicles. There are three ossicles (malleus, incus, stapes), 5–9mm in length, in each middle ear. The largest mature bones are the femurs (thigh bones), which may be longer than 45cm (Williams et al 1995). There are four general shape categories: long bones, short bones, flat bones and irregular bones. Some bones fit into more than one category; for example, the auditory ossicles and the bones of the wrist are categorized as short and irregular. Whereas there are considerable differences in the size and shape of bones, there are a number of features that are common to many bones.

Common bone features

The common features of bones (Figure 3.2) are as follows:

Articular surface: part of a bone that forms a joint with another bone.
Concave articular surface: a rounded depression.
Convex articular surface: a rounded elevation.
Facet: a small fairly flat articular surface. A convex facet on one bone usually articulates with a concave facet on an adjacent bone.
Condyle: a rounded projection of bone that provides the base for a rounded articular surface. A convex condyle on one bone usually articulates with a concave condyle on an adjacent bone.
Trochlea: a pulley-shaped condyle.
Fossa: an oval or circular depression or cavity that may also be an articular surface.
Notch: an oval depression that is often an articular surface. A notch may also take the form of a depressed region on the edge of a flat bone.

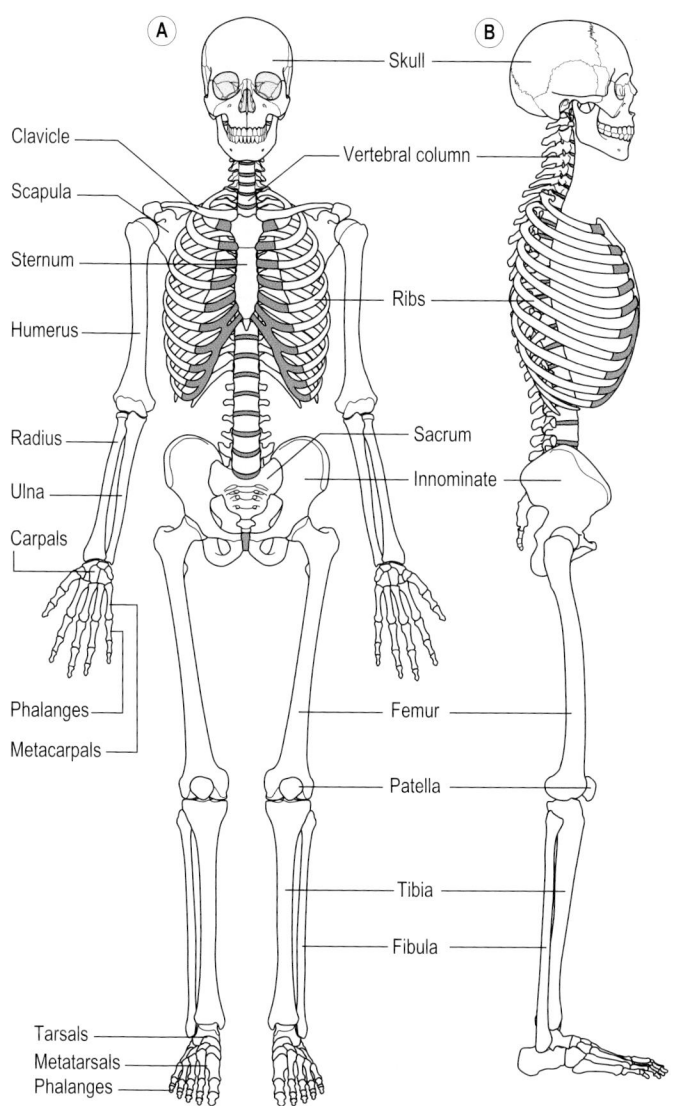

Figure 3.1 The skeleton. (A) Anterior aspect of the skeleton in the anatomical position. (B) Right lateral aspect of the skull, vertebral column and lower limb.

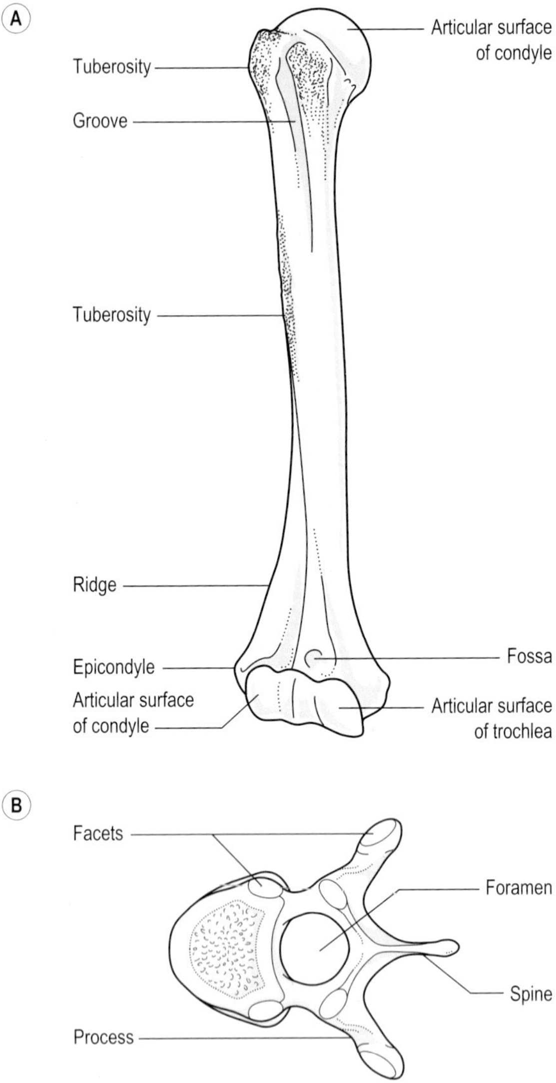

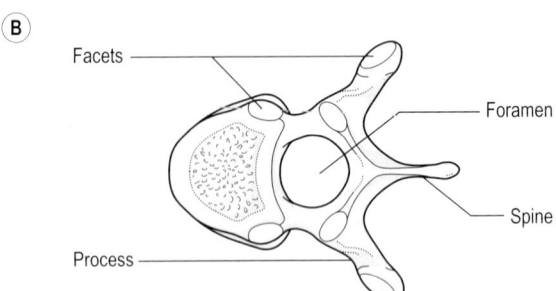

Ⓐ

Tuberosity

Groove

Tuberosity

Ridge

Epicondyle

Articular surface
of condyle

Articular surface
of condyle

Fossa

Articular surface
of trochlea

Ⓑ

Facets

Foramen

Spine

Process

Figure 3.2 Common features of bones. (A) Anterior aspect of the humerus. (B) Superior aspect of a thoracic vertebra.

Groove or sulcus: an elongated depression (like a trench). One or more tendons usually occupy grooves.

Ridge or line: an elongated elevation. A ridge is usually the site of attachment of one or more aponeuroses.

Crest: a broad ridge.

Process: a projection of bone from the main body usually providing attachment for tendons or ligaments.

Spine: a smooth process that may be slender or flat.

Epicondyle: a small process adjacent to a condyle.

Tubercle: a small roughened process.

Tuberosity: a large roughened process.

Trochanter: another name for a tuberosity used specifically in the description of the thigh bone (femur).

Foramen: a hole through a bone for the passage of blood vessels and nerves.

Those parts of the surface of a bone that do not have any of the above specific features are normally fairly smooth. These smooth areas, usually fairly large, are where muscles attach directly to the bone. When a muscle is attached to a bone by a tendon or aponeurosis, the site of attachment to the bone is usually rough and, as such, is likely to be referred to as a tubercle, tuberosity, trochanter, ridge, line or crest.

Anatomical frame of reference and spatial terminology

In order to describe the spatial orientation of the particular features of a bone, or to describe the position of one bone (or body part) in relation to another, it is necessary to use standard terminology with reference to a standard body posture. In the standard posture, also called the anatomical position (see **Figure 3.1A**), the body is upright with the arms by the sides and palms of the hands facing forward. In relation to the anatomical position, the generally accepted frame of reference, referred to as the anatomical, relative or cardinal frame of reference, describes three principal planes (median, coronal, transverse) and three principal axes (anteroposterior, vertical, mediolateral) (**Figure 3.3**). The three planes are perpendicular to each other and the three axes are perpendicular to each other.

The median plane is a vertical plane that divides the body down the middle into more or less symmetrical left and right portions. The median plane is also frequently referred to as the sagittal plane; the terms sagittal, paramedian and parasagittal (para = beside or against) are also sometimes used to refer to any plane parallel to the median plane. In this book, the term sagittal is used to refer to any plane parallel to the median plane. The mediolateral axis is perpendicular to the median plane. The

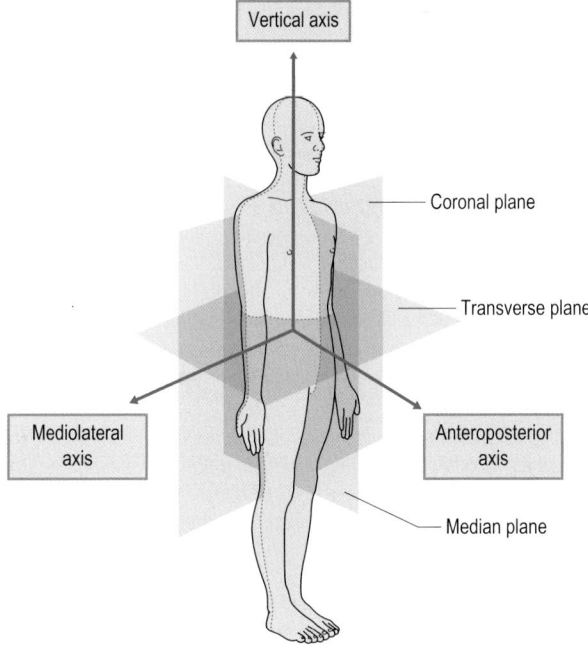

Figure 3.3 Anatomical frame of reference.

terms lateral and medial are used to describe the position of structures with respect to the mediolateral axis. Lateral means further away from the median plane and medial means closer to the median plane. For example, in the anatomical position, the lateral end of the clavicle (collarbone) articulates with the scapula (shoulder blade) and the medial end of the clavicle articulates with the sternum (breast bone). Similarly, in the anatomical position, the fingers of each hand are medial to the thumbs and the thumbs are lateral to the fingers.

The coronal plane (or frontal plane) is a vertical plane perpendicular to the median plane that divides the body into anterior and posterior portions. The anteroposterior axis is perpendicular to the coronal plane. The terms anterior (in front of) and posterior (behind) are used to describe the position of structures with respect to the anteroposterior axis. For example, the face forms the anterior part of the skull, the sternum is anterior to the vertebral column and the patella (knee cap) is anterior to the lower end of the femur. Similarly, the toes of each foot are anterior to the heels and the heels are posterior to the toes. The terms ventral and dorsal are synonymous with anterior and posterior, respectively.

The transverse plane is a horizontal plane, perpendicular to both the median and coronal planes, that divides the body into upper and lower portions. The vertical axis is perpendicular to the transverse plane. The terms superior (above) and inferior (below) are used to describe the position of structures with respect to the vertical axis. For example, as seen in **Figure 3.1**, the ribs are superior to the innominate bones (hip bones) and the patellae are inferior to the innominate bones. Similarly, the superior end of the right femur articulates with the right innominate bone to form the right hip joint, and the inferior end of the right femur articulates with the right patella and right tibia to form the right knee joint.

To describe the precise location and orientation of specific features of a particular bone, it is usually necessary to use combinations of the six spatial terms that are applicable to all bones: lateral, medial, anterior, posterior, superior and inferior. For example, a particular feature may be described as being at the anterior inferior lateral aspect of a bone; another feature may be described as being at the posterior superior medial aspect of the bone. However, there are some spatial terms that apply to some bones, but not to others. For example, the terms proximal and distal are normally only used in reference to the long bones of the limbs. Superior features of these bones (with respect to the anatomical position) are referred to as proximal, whereas inferior features of the bone are referred to as distal. For example, in each arm the proximal end of the humerus articulates with its corresponding scapula to form the shoulder joint. The distal end of the humerus articulates with the proximal ends of the radius and ulna to form the elbow joint. The distal ends of the radius and ulna articulate with the carpals to form the wrist joint.

The names of the three reference planes are often used to describe sectional views of bones. For example, **Figure 3.4** shows a coronal section through the right elbow joint. The term longitudinal section normally refers to a vertical section, as in **Figure 3.4**. A longitudinal section may be in the median plane, a paramedian plane, a coronal plane or some other vertical plane. The term cross section is a general term that may refer to a section in one of the reference planes or to an oblique plane (relative to the reference planes).

The axial skeleton

The axial skeleton consists of the skull, the vertebral column and the rib cage. The skull consists of 29 fairly flat or irregular bones that encase

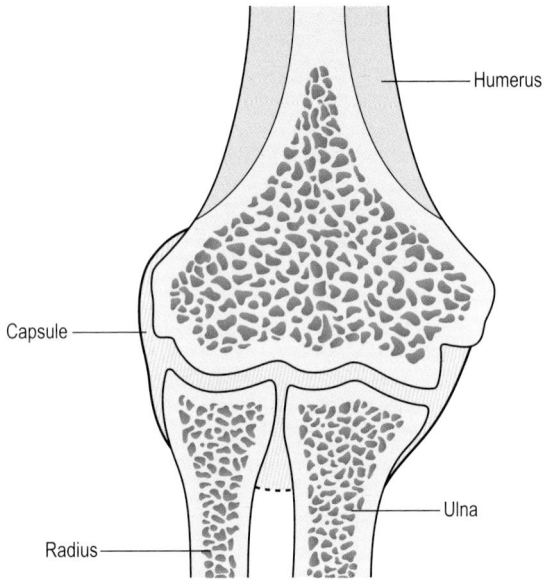

Figure 3.4 Coronal section through the right elbow joint with the arm in the anatomical position.

the brain, provide bases for the major sense organs, and form the upper and lower jaws. The vertebral column is composed of 26 irregular bones stacked on top of each other that form a curved structure in the median plane and a linear structure in the coronal plane (see **Figure 3.1**). The vertebral column supports the weight of the head, arms and trunk, and provides protection for the spinal cord. The rib cage consists of 25 bones, the sternum and 12 pairs of ribs. The sternum is a fairly flat bone. Even though the ribs are considerably curved, they are fairly flat in cross section. The rib cage is a flexible structure that provides protection for the heart and lungs; it is also very important in the ventilation of the lungs during breathing.

The skull

The 29 bones of the skull comprise 8 cranial bones (cranium), 13 facial bones (face), 6 auditory ossicles, the mandible (lower jaw) and the hyoid bone (part of the larynx). The bones of the cranium and face form a single

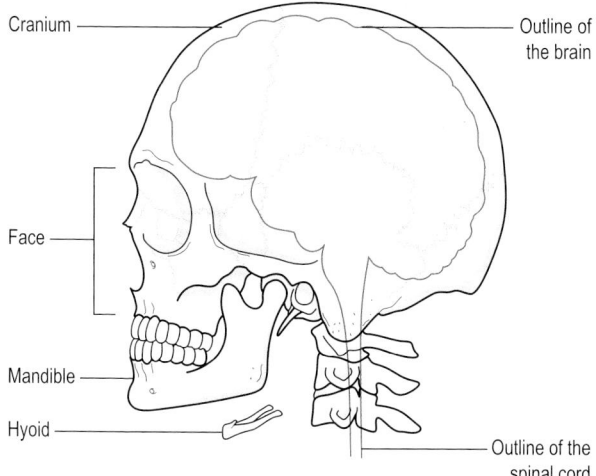

Figure 3.5 The skull in relation to the brain and spinal cord.

unit that makes up most of the skull (Figure 3.5). The cranium encloses the brain and is made up of eight relatively flat irregular bones.

The frontal bone forms the anterior and anterior superior part of the cranium including the forehead (Figure 3.6). The two parietal bones form a large part of the superior and lateral aspects of the cranium. The two temporal bones form a large part of the superior and lateral aspects of the base and sides of the cranium. The sphenoid bone together with the ethmoid bone and the inferior aspects of the frontal bone forms the anterior half of the base of the cranium. The occipital bone forms the posterior inferior aspect of the cranium and the major portion of the posterior half of the base of the cranium. The occipital bone has a large hole, the foramen magnum, situated anteriorly. The foramen is occupied by the start of the spinal cord, which is continuous with the brain (see Figure 3.5).

The 13 bones of the face form the middle third of the anterior aspect of the skull (see Figure 3.5). The facial bones form the upper jaw, the anterior part of the nasal cavity and the inferior two-thirds of the eye sockets (orbits). The upper jaw, which provides sockets for the upper teeth, is formed almost entirely by the two maxilla bones that join anteriorly in the median plane.

The mandible, or lower jaw, consists of two L-shaped plates of bone that join anteriorly in the median plane. The upright part of each half of the mandible is called the ramus (branch) and the horizontal part, called

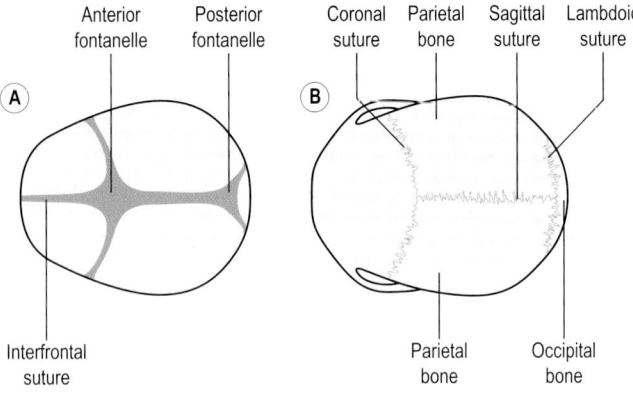

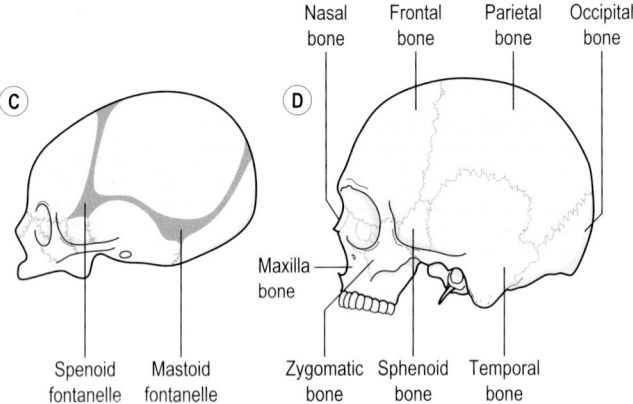

Figure 3.6 Superior and lateral aspects of the cranium and face. (A, C) Infant. (B, D) Adult.

the body, provides sockets for the lower teeth. At the posterior superior aspect of each ramus there is a convex condyle that articulates with the mandibular fossa on its corresponding temporal bone to form the corresponding temporomandibular joint (see Figure 3.5). These joints enable the mandible to swing up and down, as in closing and opening the mouth, and to move from side to side. Chewing food involves a combination of these two types of movement.

On each side of the skull, the auditory ossicles are located in a chamber, the middle ear, within the corresponding temporal bone. The three ossicles link the lateral and medial walls of the middle ear and transmit sound waves from the outer ear to the sound receptors in the inner ear.

The hyoid bone is not part of the skull, but it is convenient to describe it in relation to the skull. The hyoid is a U-shaped bone suspended in front of the neck (in front of the fourth cervical vertebra) by ligaments from the styloid processes of the temporal bones (see **Figure** 3.5). The hyoid forms part of the larynx (voice box) and provides attachment for some of the muscles that move the mouth and the tongue.

Sutures and fontanels

In the adult skull, the edges of the bones are serrated so that the bones interlock closely with each other to form immovable joints. The serrated line of the joints is similar in appearance to a line of stitches and, for this reason, each joint is called a suture (see **Figure** 3.6). In an infant, the joints between the bones of the cranium are also called sutures, even though the bones are joined by sheets of fibrous tissue and, consequently, do not interlock with each other (see **Figure** 3.6). Fibrous tissue, described in more detail in Chapter 4 (Connective tissues), is flexible, strong and, in an infant, moderately elastic.

At each of the angles (or corners) of the parietal bones, the fibrous tissue is in the form of a small sheet, called a fontanel (Williams et al 1995). Since each parietal bone has four angles and the parietal bones join each other superiorly in the median plane, there are six fontanels. The anterior fontanel, which normally closes within the first 18 months, is at the junction of the parietal and frontal bones. At birth the frontal bone is in two halves, joined by the interfrontal or metopic suture; these halves normally fuse together within the first 2 years. The posterior fontanel is at the junction of the parietal and occipital bones and normally closes within the first 2 months. There are two sphenoid fontanels, one on each side of the skull at the junction of the parietal, frontal, sphenoid and temporal bones. The sphenoid fontanels normally close within the first 3 months. There are also two mastoid fontanels, one on each side of the skull at the junction of the parietal, occipital and temporal bones. The mastoid fontanels normally close within the first 2 years.

Key Concepts

The skull consists of 29 fairly flat or irregular bones. The bones of the cranium (8 bones), face (13 bones) and the auditory ossicles (3 in each middle ear) form a single unit that makes up most of the skull. The other two bones are the mandible and the hyoid

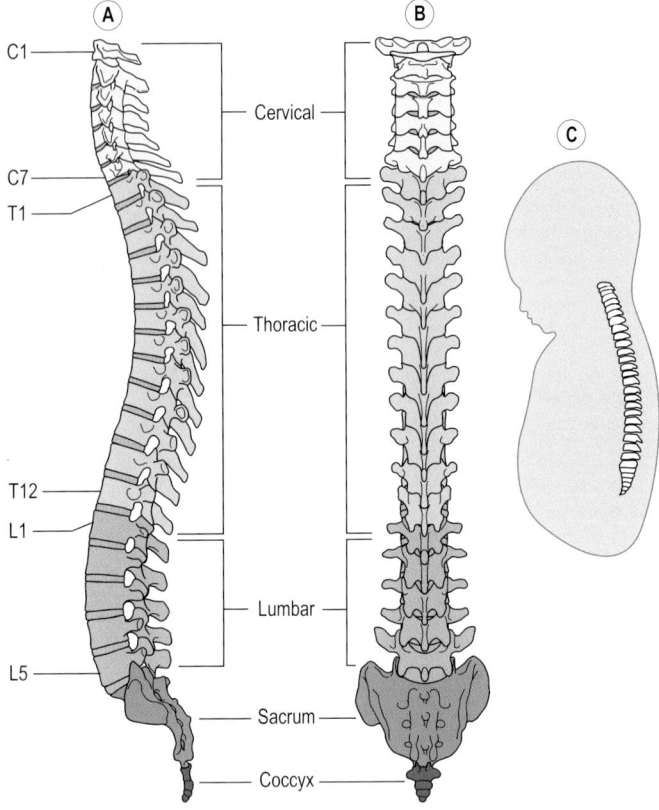

C1
C7
T1
T12
L1
L5

Cervical
Thoracic
Lumbar
Sacrum
Coccyx

Figure 3.7 The vertebral column. (A) Left lateral aspect in an adult. (B) Posterior aspect in an adult. (C) Left lateral aspect in an infant.

The vertebral column

Prior to maturity, the vertebral column consists of 33 or 34 irregular bones called vertebrae. The vertebrae are divided into five fairly distinct groups: cervical, thoracic, lumbar, sacral and coccygeal (**Figure 3.7**). The neck consists of seven cervical vertebrae. The thoracic or chest region consists of 12 thoracic vertebrae that provide articulation for the 12 pairs of ribs. The lower back consists of five lumbar vertebrae. The five sacral vertebrae form the posterior part of the pelvis; at maturity the sacral vertebrae fuse together to form the sacrum. The four or five coccygeal vertebrae are small and represent a vestigial tail. The coccygeal vertebrae normally fuse together at maturity to form the coccyx, or tailbone, which is approximately 3 cm long and is attached to the sacrum by ligaments.

When viewed from the side, the whole of the vertebral column of a newborn infant is concave anteriorly (see **Figure 3.7C**). Between 3 and 6 months of age the child learns to hold his head upright and, as a result, the shape of the cervical region changes from concave anteriorly to convex anteriorly. Similarly, as the child learns to stand and walk, between 10 and 18 months of age, the shape of the lumbar region also changes from concave anteriorly to convex anteriorly. The cervical and lumbar curves are referred to as secondary curves as they develop as the child adopts an upright posture. The thoracic and sacrococcygeal curves are called primary curves as they are concave anteriorly throughout life (see **Figure 3.7**).

Structure of a vertebra

At birth, each vertebra, with the exception of the first two cervical vertebrae, consists of three bony elements united by cartilage (Williams et al 1995) (**Figure 3.8A**). The anterior element, the centrum or body, is basically a block of bone with slightly concave (waisted) sides and fairly flat kidney-shaped superior and inferior surfaces. The bodies of the vertebrae are mainly responsible for transmitting loads, especially the weight of the head, arms and trunk. The posterior elements are curved struts that form the two halves of an arch, that is, the vertebral or neural arch. Each half of the arch consists of an anterior portion called the pedicle and a posterior portion called the lamina. The two laminae normally fuse together posteriorly during the first year (see **Figure 3.8B**). The pedicles normally fuse with the lateral superior posterior aspects of the body of the vertebra between the third and sixth years (see **Figure 3.8C**). The hole formed by the arch and the posterior aspect of the body is called the vertebral foramen. In the vertebral column, the spinal cord passes through all of the vertebral foramina.

After fusion of the laminae, seven processes arise from the arch. The spine of the vertebra extends backward from the point of fusion of the laminae. On each side of the arch, three processes arise from the junction of the pedicle and lamina. A transverse process extends laterally, a superior articular process extends upward and an inferior articular process extends downward (**Figure 3.9**). The spine and transverse processes are levers that provide areas of attachment for muscles, tendons and ligaments. With respect to each pair of adjacent vertebrae, the superior articular processes of the lower vertebra articulate by means of facets with the inferior articular processes of the upper vertebra (see **Figure 3.9B**

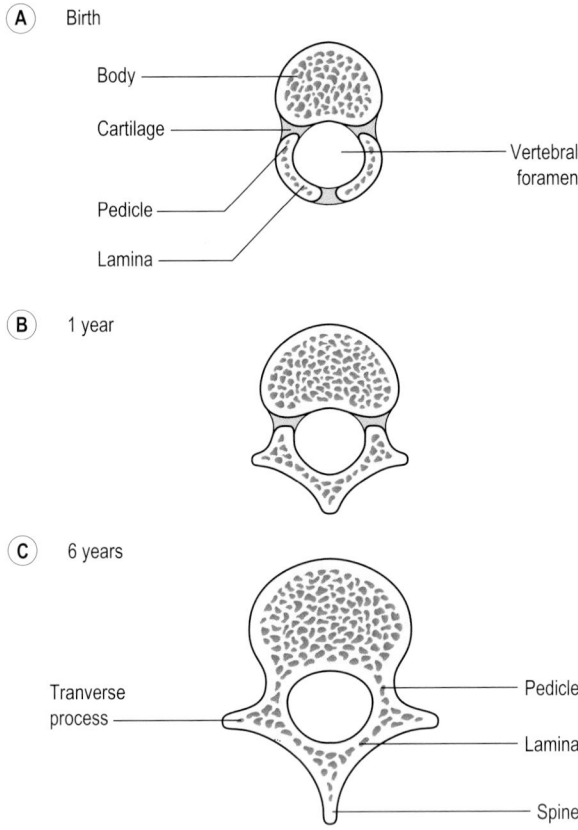

(A) Birth

Body

Cartilage

Vertebral foramen

Pedicle

Lamina

(B) 1 year

(C) 6 years

Tranverse process

Pedicle

Lamina

Spine

Figure 3.8 Early stages in the development of a vertebra (superior aspect).

and **Figure 3.10**). These joints are called facet joints or apophyseal joints. In most upright postures, the facet joints transmit some load. In general, the load transmitted by these joints decreases with flexion of the trunk (bending forward) and increases with extension of the trunk (bending backward).

The bodies of each pair of adjacent vertebrae, except the first two cervical vertebrae, are joined by a tough rubbery kidney-shaped disc of fibrocartilage called an intervertebral disc to form an intervertebral joint. Consequently, whereas the first two cervical vertebrae are joined by two

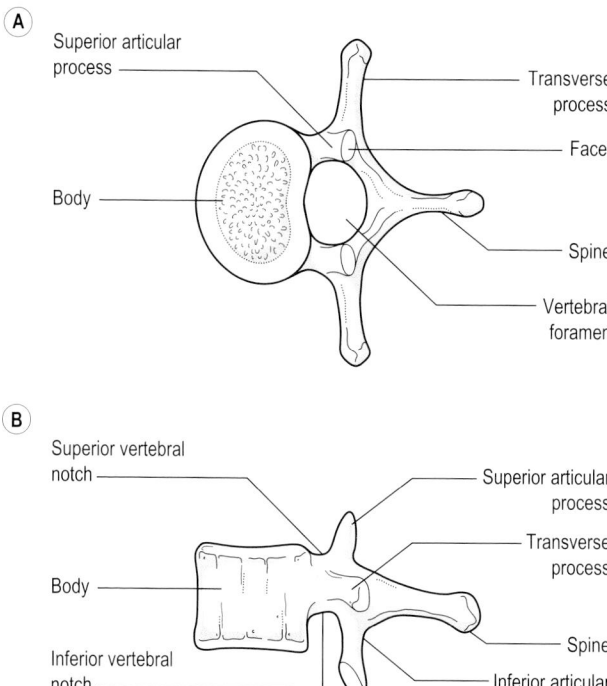

Figure 3.9 A typical vertebra. (A) Superior aspect. (B) Left lateral aspect.

facet joints, the other vertebrae are joined by one intervertebral joint and two facet joints. The type and range of movement between adjacent vertebrae is largely determined by the thickness of the intervertebral disc (which deforms in response to loading, rather like a pencil eraser) and the orientation of the facet joints (whose articular surfaces slide on each other). The different types of joints are described in Chapter 5 (The articular system).

On each side of the vertebral arch is a depression in the superior aspect of the pedicle called the superior vertebral notch (see **Figure 3.9B**). Since the pedicle joins the posterior superior aspect of the body, there is a much larger inferior vertebral notch beneath the pedicle. With respect to each pair of adjacent vertebrae, the inferior notch of the upper

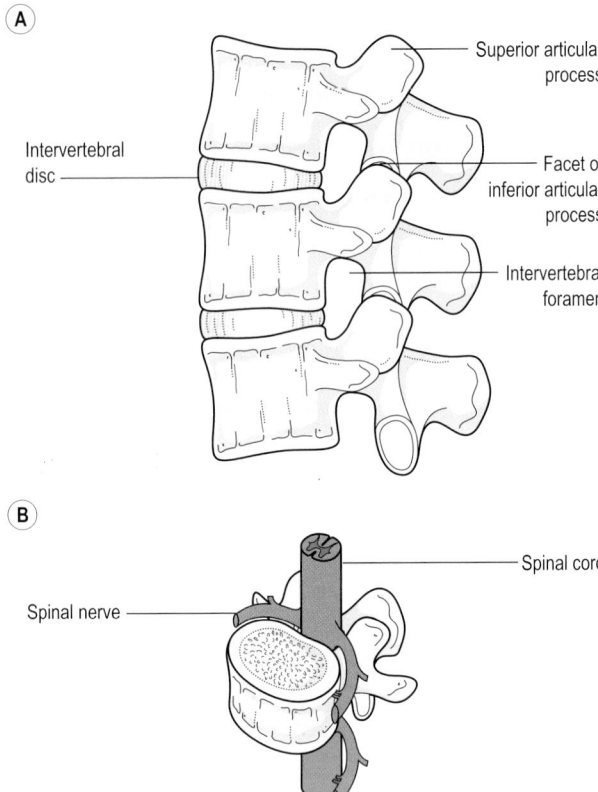

Figure 3.10 (A) Left lateral aspect of three articulated lumbar vertebrae. (B) Relationship of the spinal cord and spinal nerves to a lumbar vertebra.

vertebra and the superior vertebral notch of the lower vertebra form a hole called the intervertebral foramen (**Figure 3.10A**). A spinal (peripheral) nerve occupies the intervertebral foramen (**Figure 3.10B**).

Distinguishing features of vertebrae

Whereas the vertebrae all have the same basic structure, there are differences between the different regions of the vertebral column in the size and shape of the bodies and in the size, shape and orientation of the processes and facet joints. The bodies and processes of the vertebrae gradually increase in size from the second cervical vertebra down

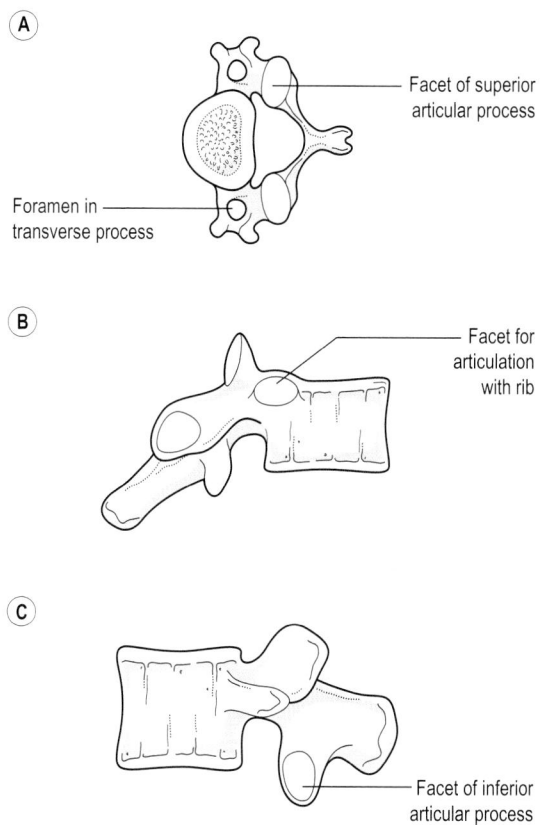

(A)

Facet of superior
articular process

Foramen in
transverse process

(B)

Facet for
articulation
with rib

(C)

Facet of inferior
articular process

Figure 3.11 (A) Superior aspect of a cervical vertebra. (B) Right lateral aspect of a thoracic vertebra. (C) Left lateral aspect of a lumbar vertebra.

to the sacrum; this increase reflects the gradual increase in weight that the vertebrae have to support (see **Figure** 3.7A). The changes in the size, shape and orientation of the processes and facet joints are fairly gradual within each of the cervical, thoracic and lumbar regions, but tend to be more marked at the junctions between the regions (see **Figure** 3.7A). In the mature vertebral column, the cervical, thoracic and lumbar vertebrae have characteristics that distinguish them from each other.

Cervical All cervical vertebrae (C1 to C7) have a hole called a transverse foramen in each of their transverse processes (**Figure** 3.11A). Only cervical vertebrae have this characteristic. The facets of the superior and inferior articular processes of the cervical vertebrae articulate in oblique planes that slope downward laterally and posteriorly. The orientation of the facet joints, the short transverse processes, the relatively thick intervertebral discs and the relatively short spines of C3 to C6 all combine to give a fairly large range of movement in the cervical region as a whole compared to other regions of the column.

Thoracic Thoracic vertebrae (T1 to T12) can be identified by the presence of facets on the lateral aspects of the bodies for articulation with the heads (posterior ends) of the ribs (see **Figure 3.11B**). The spines of the thoracic vertebrae are fairly long and tend to closely overlap each other, especially in the middle of the region (see **Figure 3.7A**). The transverse processes of the thoracic vertebrae are also fairly long; they gradually decrease in length from T1 to T12. The superior and inferior articular facets articulate in a plane that slopes sharply downward posteriorly. The overlapping spines, relatively thin intervertebral discs and splinting effect of the ribs result in a smaller overall range of movement in the thoracic region than in the cervical region.

Lumbar The lumbar vertebrae (L1 to L5) have fairly long transverse processes and large, flat, rectangular-shaped spines (see **Figure 3.11C**). The main distinguishing feature of the lumbar vertebrae is the orientation of the facets on the superior and inferior articular processes. The facets on the superior articular processes face medially and posteriorly, and the facets on the inferior articular processes face laterally and anteriorly. The orientation of the facet joints severely limits rotation of the lumbar vertebrae about a vertical axis. However, the relatively thick intervertebral discs in the lumbar region ensure a much greater range of movement in other directions.

Sacrum The sacral vertebrae (S1 to S5) become progressively smaller from S1 through S5. The sacrum is formed by the fusion or partial fusion of the sacral vertebrae. When viewed from the front (or the back) the sacrum is more or less triangular with the apex pointing downward (**Figure 3.12A**). The anterior edge of the upper surface of the first sacral vertebra projects forward and is called the sacral promontory. The anterior aspect of the sacrum is concave, largely due to the orientation of S3, S4 and S5 (**Figure 3.12B**). In the anatomical position, the large upper portion of the sacrum (S1 and S2) is tilted forward, which tends to accentuate the lumbar curve (see **Figure 3.7A**).

On each superior lateral aspect of the sacrum is a fairly large C- or L-shaped articular surface called the auricular surface (auricle = ear-shaped) (see **Figure 3.12B**). The auricular surfaces are formed by the lateral expansions of the fused transverse processes of S1, S2 and S3. The auricular surfaces of the sacrum articulate with the hip bones (innominate

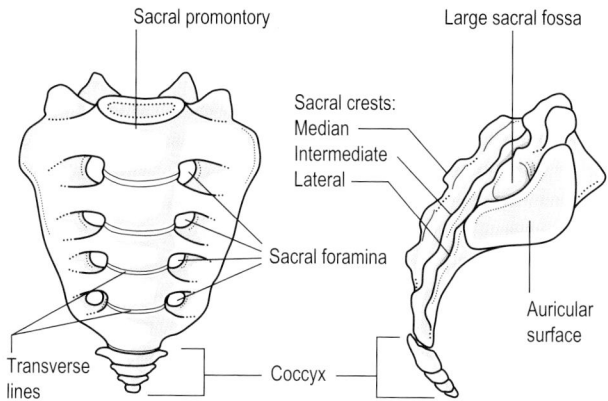

Sacral promontory

Large sacral fossa

Sacral crests:
Median
Intermediate
Lateral

Sacral foramina

Auricular
surface

Transverse
lines

Coccyx

Figure 3.12 (A) Anterior aspect of the sacrum. (B) Right lateral aspect of the sacrum.

bones) to form the sacroiliac joints (see **Figure 3.1**). The sacrum and the left and right hip bones form a complete bony ring called the pelvis or pelvic girdle. Consequently, the sacrum is an important part of the vertebral column and the pelvis.

Key *Concepts*

Whereas the vertebrae all have the same basic structure, there are differences between the different regions of the vertebral column in the size and shape of the bodies and in the size, shape and orientation of the processes and facet joints

The rib cage

The rib cage, which consists of 12 pairs of ribs and the sternum, provides protection for the heart and lungs and facilitates breathing. The sternum is a fairly flat bone and the ribs are also fairly flat in cross section. As the ribs form the major part of the rib cage, they have a distinct curved shape. The heads (posterior ends) of the ribs articulate with the thoracic vertebrae. The anterior ends of the upper 10 pairs of ribs are attached to the sternum by pieces of cartilage called costal cartilages (costa = rib). The upper seven pairs of ribs are attached to the sternum by separate costal cartilages and are therefore sometimes referred to as true ribs. The costal cartilages of the eighth, ninth and tenth pairs of ribs fuse with each other before fusing with the costal cartilages of the

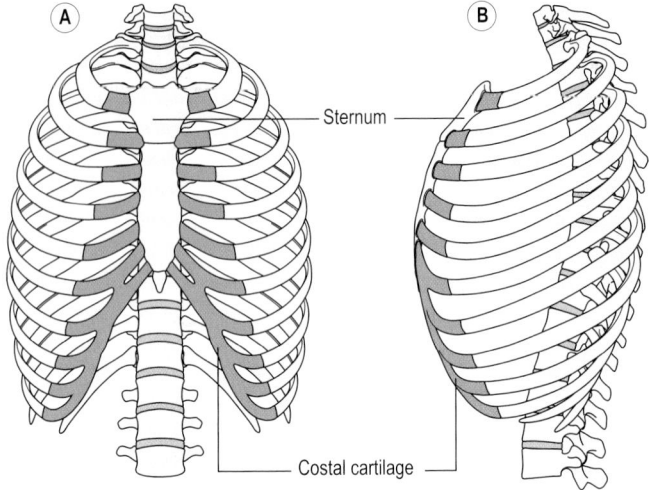

Figure 3.13 (A) Anterior aspect of the rib cage. (B) Left lateral aspect of the rib cage.

seventh ribs (**Figure 3.13**). Consequently, whereas the upper seven pairs of ribs have direct cartilaginous attachments to the sternum, the eighth, ninth and tenth pairs of ribs have an indirect cartilaginous attachment to the sternum. The lower two pairs of ribs do not attach onto the sternum; the anterior ends of these ribs are free, and they are referred to as floating ribs. Because none of the lower five pairs of ribs has a direct cartilaginous or other attachment to the sternum, these ribs are sometimes referred to as false ribs.

Movement of the ribs

The intercostal spaces (spaces between the ribs) are largely occupied by muscles (intercostal muscles), which, in association with other muscles of the thorax, move the ribs during breathing. During inspiration (breathing in) the ribs and sternum swing upward and outward which decreases the pressure inside the thorax and simultaneously drives air into the lungs. During expiration (breathing out) the ribs and sternum are pulled back down by the elasticity of the surrounding soft tissues (costal cartilages, ligaments and muscles that are stretched during inspiration). The downward movement of the ribs and sternum increases the pressure inside the thorax and simultaneously drives air out of the lungs.

Key Concepts

The rib cage consists of the sternum and 12 pairs of ribs. It is a fairly flexible structure that provides protection for the heart and lungs and, in association with intercostal muscles, facilitates breathing

The appendicular skeleton

The appendicular skeleton (126 bones) consists of the bones of the upper and lower limbs. In an adult, each upper limb consists of 32 bones and each lower limb consists of 31 bones.

The upper limb

There are five regions in each upper limb: shoulder (scapula and clavicle), upper arm (humerus), lower arm (radius and ulna), wrist (8 carpals) and hand (5 metacarpals and 14 phalanges).

Shoulder

The shoulder region of each upper limb consists of a scapula (shoulder blade) and a clavicle (collarbone) (Figure 3.14). Together with the sternum, the scapulae and clavicles of both upper limbs form an incomplete ring of bone called the shoulder girdle (see Figure 3.1). The arms are suspended from the shoulder girdle. The medial end of each clavicle articulates with the sternum to form a sternoclavicular joint, and the lateral end of each clavicle articulates with the acromion process of the corresponding scapula to form an acromioclavicular joint (see Figure 3.1). The scapulae are not joined to the axial skeleton, but are held in position at the lateral superior posterior aspects of the rib cage by muscles. Consequently, each scapula has a considerable range of movement. Most movements of the shoulder region involve movements at the sternoclavicular and acromioclavicular joints.

Upper arm

The humerus, the only bone in the upper arm, is a typical long bone consisting of a relatively long shaft between two fairly bulbous ends (Figure 3.15). The proximal end of the humerus is dominated by the head which is an almost perfect hemisphere and articulates with the glenoid fossa to form the shoulder joint. The distal end of the humerus has a cylindrical-shaped articular surface consisting of two condyles, the capitulum and trochlea, fused together side by side.

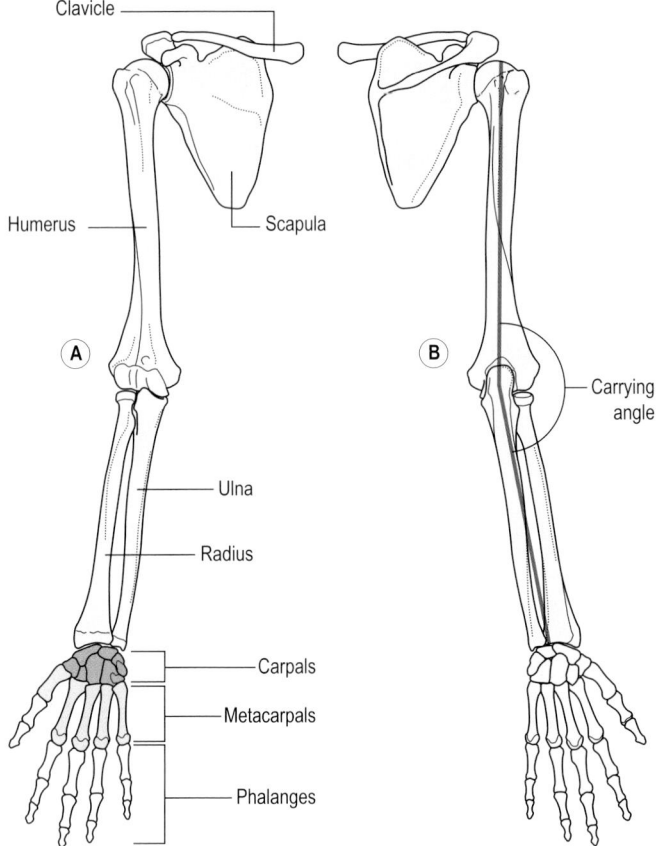

Clavicle

Humerus

Scapula

Ⓐ

Ⓑ

Carrying angle

Ulna

Radius

Carpals

Metacarpals

Phalanges

Figure 3.14 (A) Anterior aspect of the right upper limb. (B) Posterior aspect of the right upper limb.

Lower arm

The lower arm consists of two long bones, the radius and the ulna. In the anatomical position, the radius is lateral to the ulna (**Figure 3.16**). The proximal end of the ulna is dominated by a large pulley-shaped concave articular surface, the trochlea notch, which articulates with the trochlea of the humerus to form part of the elbow joint. The proximal end of the radius consists of a drum-shaped head separated from the main part of

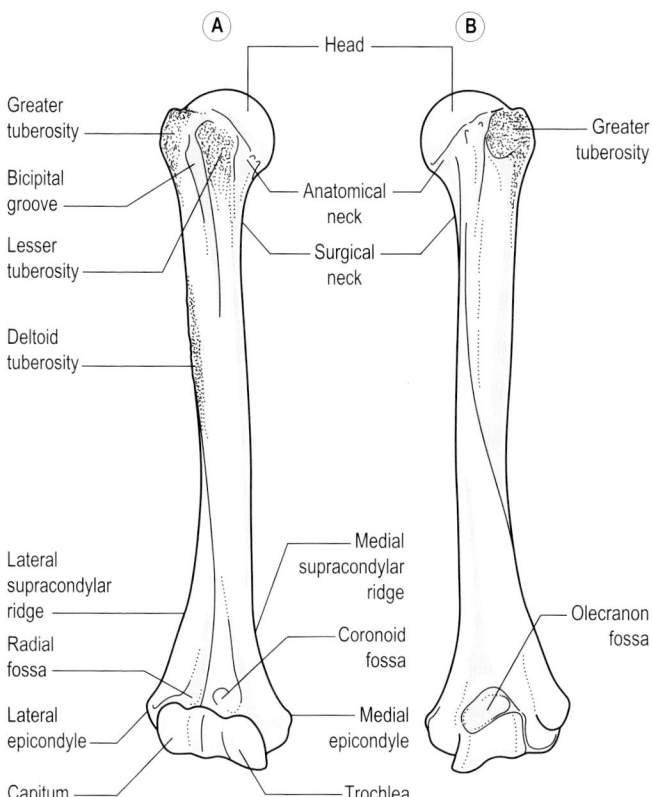

Greater
tuberosity

Bicipital
groove

Lesser
tuberosity

Deltoid
tuberosity

Lateral
supracondylar
ridge

Radial
fossa

Lateral
epicondyle

Capitum

Head

Anatomical
neck

Surgical
neck

Greater
tuberosity

Medial
supracondylar
ridge

Coronoid
fossa

Medial
epicondyle

Trochlea

Olecranon
fossa

Figure 3.15 The right humerus. (A) Anterior aspect. (B) Posterior aspect.

the shaft by a short cylindrical neck (see **Figure 3.16**). The circular side of the head and the superior surface of the head form a continuous articular surface. The side articulates with the radial notch on the ulna, and the superior surface articulates with the capitulum on the humerus. The elbow joint consists of the joints between the trochlea and trochlea notch, and the capitulum and head of the radius.

The distal end of the ulna has a small drum-shaped head with a small projection on its posteromedial aspect called the styloid process of the ulna. The lateral part of the distal end of the radius forms a small projection called the styloid process of the radius. The inferior aspect of the distal end

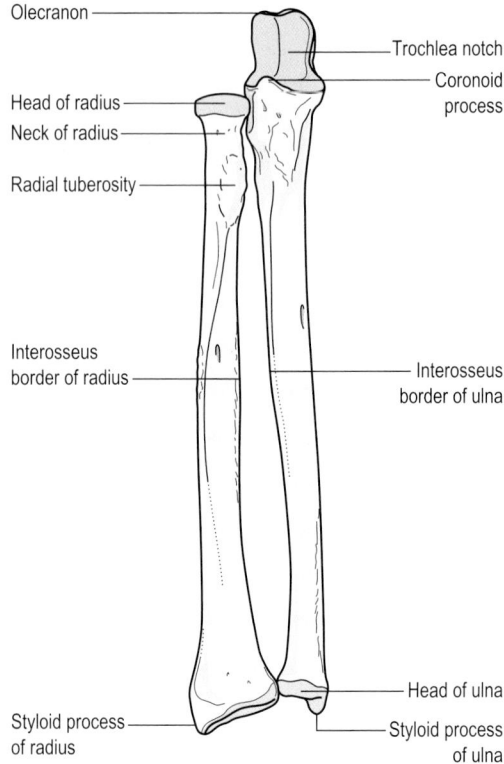

Olecranon

Trochlea notch

Coronoid process

Head of radius

Neck of radius

Radial tuberosity

Interosseus border of radius

Interosseus border of ulna

Head of ulna

Styloid process of radius

Styloid process of ulna

Figure 3.16 Anterior aspect of the right ulna and radius.

is dominated by a fairly large more or less quadrangular concave articular surface that forms part of the wrist joint. Adjacent to and continuous with the medial edge of this surface is the ulnar notch, a small articular surface. This notch articulates with the side of the head of the ulna.

Wrist and hand

The wrist consists of eight small irregular bones called carpals that articulate with each other to form the carpus (Figure 3.17). The carpals are arranged in a proximal and distal row. The proximal row articulates with the distal ends of the radius and ulna to form the wrist joint. The series of joints between the proximal and distal rows is called the midcarpal joint.

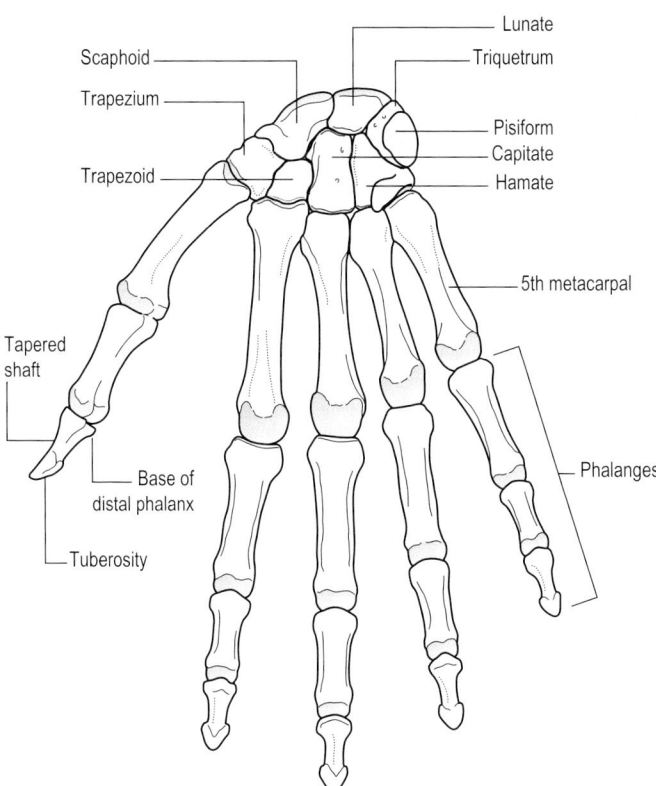

Lunate
Scaphoid
Triquetrum
Trapezium
Pisiform
Capitate
Trapezoid
Hamate

5th metacarpal

Tapered shaft

Base of distal phalanx

Phalanges

Tuberosity

Figure 3.17 Anterior aspect of the right wrist and hand.

The hand consists of 5 metacarpals and 14 phalanges (or digits) (see Figure 3.17). In life, the metacarpals are joined together by soft tissues and form the palm of the hand on the anterior aspect. The metacarpals are miniature long bones and each metacarpal consists of a base (the proximal end), a shaft and a head (the distal end). The bases of the metacarpals articulate with the distal row of carpals to form the carpometacarpal joints. The combined ranges of movement in the wrist, midcarpal, and carpometacarpal joints facilitate a large range of movement for the hand as a whole. The heads of the metacarpals articulate with the bases of the proximal phalanges to form the metacarpophalangeal joints that link the thumb and fingers to the palm of the hand. Each of the four fingers consists of three phalanges (proximal, middle and distal), whereas the thumb has only two (proximal, distal).

Key *Concepts*

Each upper limb consists of 32 bones: shoulder (scapula and clavicle), upper arm (humerus), lower arm (radius and ulna), wrist (8 carpals) and hand (5 metacarpals and 14 phalanges)

The lower limb

There are five regions in each lower limb: hip (innominate), pelvis (sacrum and right and left innominates), upper leg (femur and patella), lower leg (tibia and fibula) and foot (7 tarsals, 5 metatarsals and 14 phalanges).

Hip

Together with the sacrum, the right and left innominate bones form a complete ring of bone called the pelvis or pelvic girdle (Figure 3.18). Consequently, the innominate bones attach the legs to the axial skeleton. Each innominate bone develops from three bones called the ilium, ischium and pubis, which fuse together at maturity. The region where the three bones fuse together is dominated by a large semispherical concavity called the acetabulum, which articulates with the head of the femur to form the hip joint (Figure 3.19A).

The ilium comprises the upper two-fifths of the acetabulum and the large, more or less flat portion of the innominate bone above the acetabulum (see Figure 3.19A). The large, flat upper part of the ilium is called the wing of the ilium. The superior border of the wing of the ilium forms a broad crest called the iliac crest, which can be felt beneath the skin just above the hip joint. The iliac crest provides attachment for muscles comprising the wall of the abdomen. On the posterior medial aspect of the wing of the ilium there is a large C-shaped or L-shaped auricular surface that articulates with the auricular surface on the corresponding side of the sacrum to form the corresponding sacroiliac joint (see Figures 3.18 and 3.19). The large lateral and medial surfaces of the wing of the ilium provide attachment for muscles that move the hip joint. The medial surface of the wing also supports the contents of the abdomen.

The ischium, which forms the posterior inferior portion of the innominate bone, consists of the body and the ramus. The body is a more or less vertical pillar that transmits the weight of the trunk, head and arms to the support surface when the individual is sitting. The superior part of the body forms the posterior inferior two-fifths of the acetabulum. Below

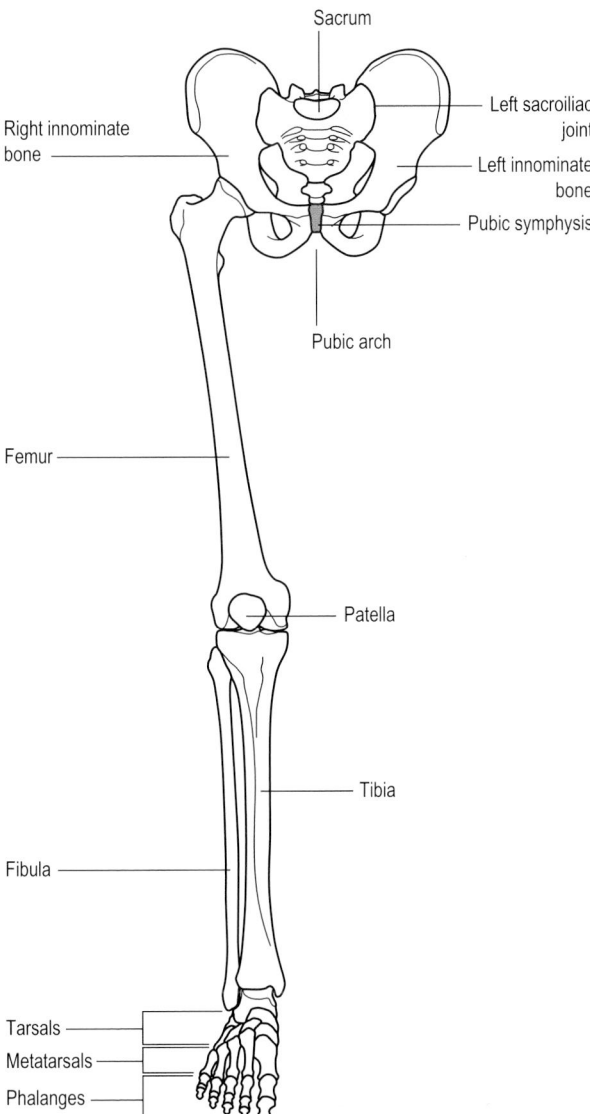

Sacrum

Left sacroiliac joint

Right innominate bone

Left innominate bone

Pubic symphysis

Pubic arch

Femur

Patella

Tibia

Fibula

Tarsals

Metatarsals

Phalanges

Figure 3.18 Anterior aspect of the pelvis and right lower limb.

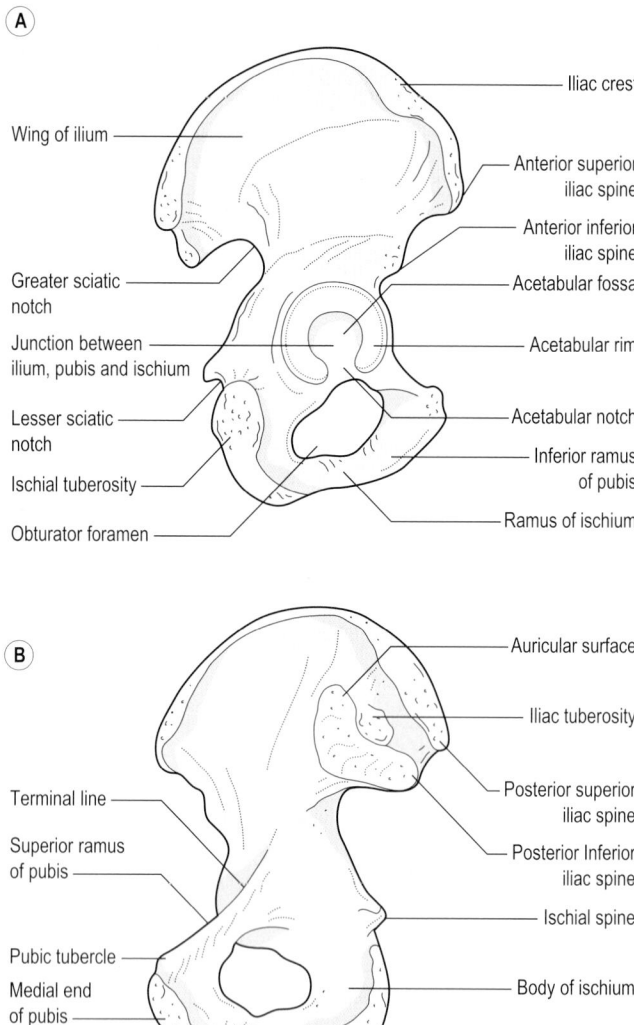

Figure 3.19 The right innominate bone. (A) Lateral aspect. (B) Medial aspect.

the acetabulum the body of the ischium is characterized by a large ischial tuberosity on its posterior inferior lateral aspect (see **Figure** 3.19A). The ramus of the ischium is a broad, flat process that arises from the base of the body and projects medially, forward and upward to articulate with the pubis.

The pubis forms the anterior inferior portion of the innominate bone. It consists of the body, superior ramus and inferior ramus. The body forms the anterior inferior one-fifth of the acetabulum. The superior ramus extends medially and also slightly forward and downward from the body to join the medial end of the inferior ramus. The junction between the two rami of the pubis forms a fairly broad, flat region. The medial surface of this junction, the medial surface of the pubis, is elliptical in shape and lies in the median plane. The long axis of the ellipse is inclined at an angle of approximately 45° to the coronal plane. The medial surfaces of the right and left pubic bones articulate in the median plane to form the pubic symphysis joint (see **Figure** 3.18). The inferior ramus of the pubis projects downward and backward laterally to join the anterior end of the ischium ramus (see **Figure** 3.19). The inverted V-shaped notch formed by the inferior borders of the right and left inferior pubic rami is called the pubic arch (see **Figure** 3.18).

The pubis and ischium are both essentially V-shaped and joined at their free ends. Consequently, when fused together, the two bones create a large foramen called the obturator foramen on account of its close proximity to the obturator nerve.

Key *Concepts*

The innominate bones link the lower limbs to the axial skeleton. Each innominate bone develops from three bones called the ilium, ischium and pubis, which fuse together at maturity

Pelvis

Pelvis is a Latin word meaning basin (due to the large wings of the ilia, which give the impression of an incomplete bowl when the pelvis is viewed from an anterior superior aspect) (**Figure** 3.20). The upper part of the pelvis (wings of the ilia and the upper two-thirds of the sacrum) provides a base of support for the upper body. The lower part of the pelvis (ischium and pubis) transmits the weight of the upper body to the legs when standing and to the chair seat when sitting. The margin between the upper and lower parts of pelvis is delineated by a continuous ridge called the inlet or pelvic brim (see **Figure** 3.20B and D).

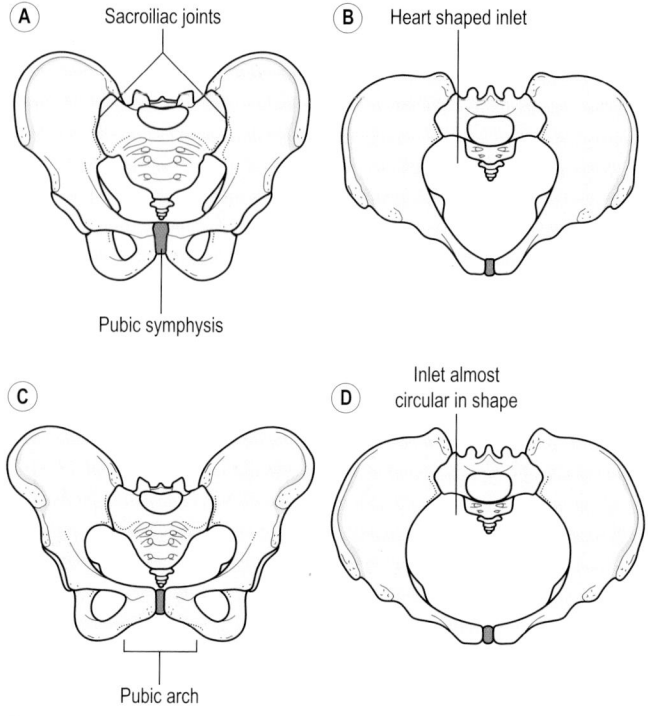

A Sacroiliac joints

Pubic symphysis

B Heart shaped inlet

C

Pubic arch

D Inlet almost circular in shape

Figure 3.20 (A) Anterior aspect of the male pelvis. (B) Anterior superior aspect of the male pelvis. (C) Anterior aspect of the female pelvis. (D) Anterior superior aspect of the female pelvis.

Whereas the pelvis is similar in structure in males and females, there are four main differences in the shape of the pelvis between males and females due, it is assumed, to the childbearing functions of the female (see **Figure 3.20**):

1. The inlet of the male pelvis is heart shaped, whereas that of the female pelvis is more circular.

2. The pubic bones are more in line with each other in the female than in the male. Consequently, the angle of the pubic arch is obtuse in the female and acute in the male.

3. The relative distance between the acetabulums is greater in the female than in the male. This results in a relatively greater girth around the hips in the female compared to the male.

4. In the male, the sacrum is curved such that the lower half of the sacrum and the coccyx bend forward. This curvature reduces the front to back dimension of the lower pelvis. In the female, the sacrum is relatively straight, which tends to maintain a fairly constant front to back dimension in the lower pelvis.

Upper leg

The upper leg or thigh contains a long bone called the femur and a relatively small bone called the patella (knee cap), which articulates with the distal end of the femur. The femur is the longest and strongest bone in the skeleton. The proximal end of the femur consists of a nearly spherical shaped head, which articulates with the acetabulum to form the hip joint (see **Figure 3.18** and **Figure 3.21**). The head is joined obliquely to the shaft by a thick neck that runs laterally downward, and backward from the head to the anterior medial region of the proximal end of the shaft. A large process called the greater trochanter dominates the superior posterior lateral region of the proximal end of the shaft. At the base of the neck on the posterior medial aspect of the shaft is another fairly large process called the lesser trochanter.

Like the humerus, the upper two-thirds of the shaft of the femur is cylindrical and the lower one-third gradually becomes broader (in the coronal plane) toward the distal end. In a paramedian plane, the anterior surface of the femur is slightly convex and the posterior surface is slightly concave (see **Figure 3.1B**). The distal end of the femur consists of two large convex condyles, the lateral and medial condyle, fused together side by side anteriorly. The condyles are separated posteriorly by a large notch called the intercondylar notch (or intercondylar fossa). The upper part of the common anterior portion of the articular surface is called the patellar surface. The patellar surface is pulley shaped – depressed in the middle in the sagittal plane – and articulates with the posterior surface of the patella to form the patellofemoral joint. During extension and flexion of the knee joint, the patella slides up and down on the patellar surface and condyles of the femur.

The patella is a sesamoid bone (Greek for resembling a sesame seed), i.e. a bone that is partially embedded in a tendon (**Figure 3.22**). A sesamoid bone tends to increase the moment arm and, in turn, the mechanical efficiency of the associated musculotendinous unit and prevent the tendon from rubbing on an adjacent bone. The whole of the anterior surface and the inferior quarter of the posterior surface are embedded in the quadriceps tendon. The upper three-quarters of the posterior surface articulates with the patellar surface of the femur when the knee joint is extended and with the condyles of the femur when the knee joint is

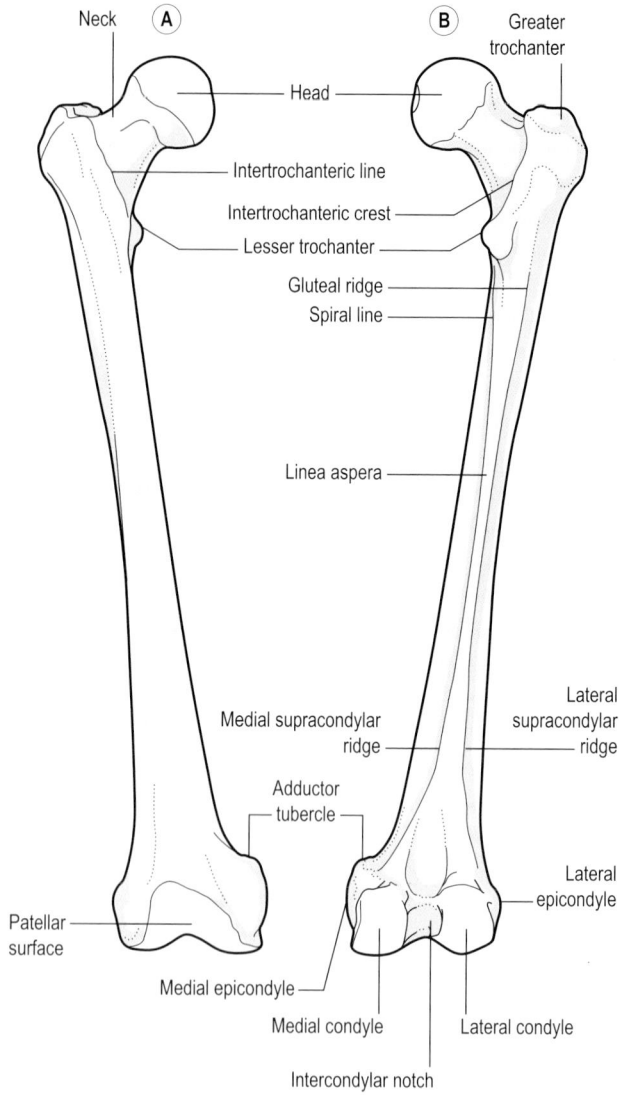

Figure 3.21 The right femur. (A) Anterior aspect. (B) Posterior aspect.

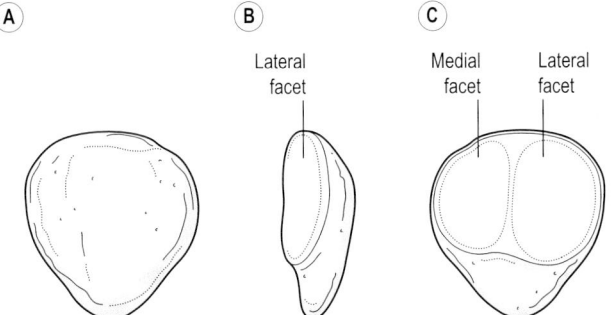

Figure 3.22 The right patella. (A) Anterior aspect. (B) Right lateral aspect. (C) Posterior aspect. The entire patella apart from the facets is embedded in the quadriceps tendon.

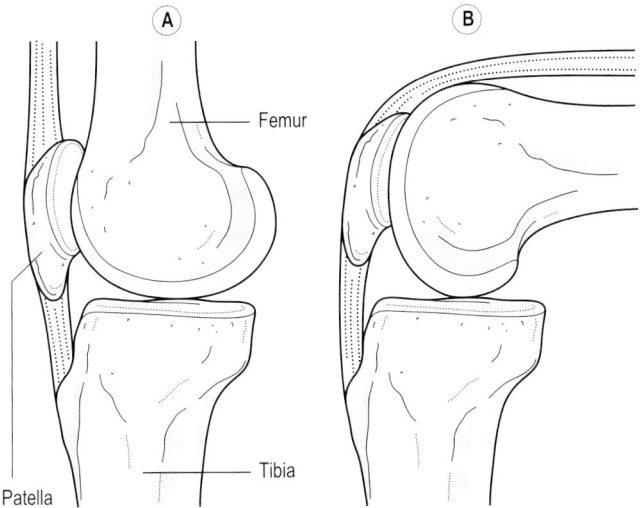

Figure 3.23 Orientation of the patella to the femur in knee extension (A) and knee flexion (B).

flexed (Figure 3.23). The patella increases the mechanical efficiency of the quadriceps muscle group and prevents the quadriceps tendon from rubbing against the patellar surface of the femur.

Lower leg

The lower leg or shank contains two long bones, the tibia and the fibula, aligned with their shafts more or less parallel to each other (Figure 3.24).

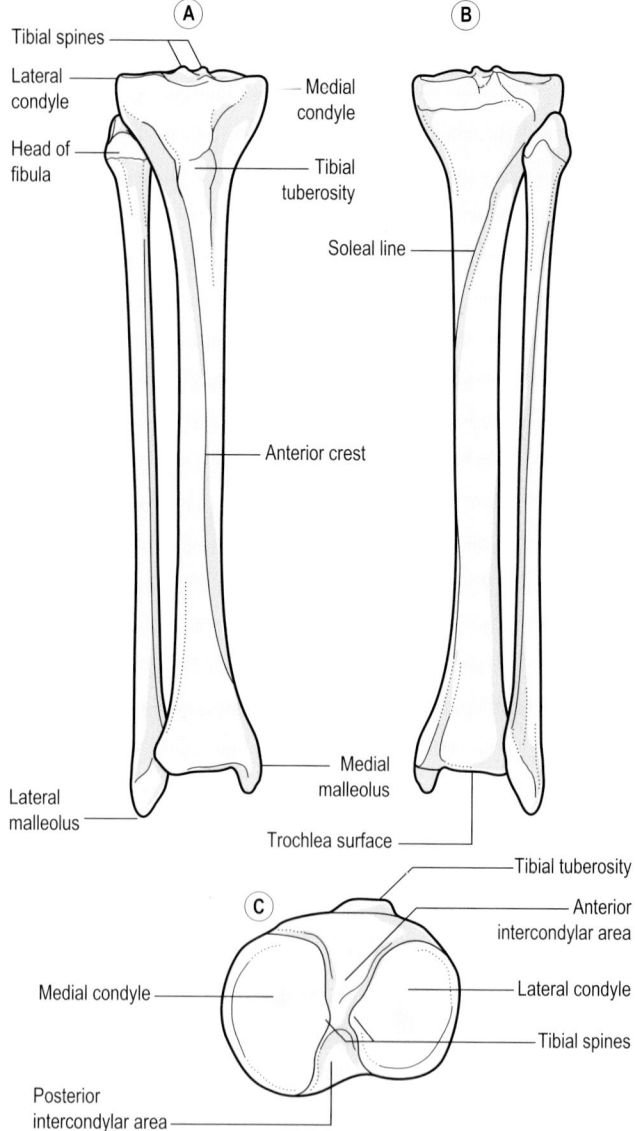

Tibial spines

Lateral condyle

Head of fibula

Mcdial condyle

Tibial tuberosity

Soleal line

Anterior crest

Medial malleolus

Lateral malleolus

Trochlea surface

Tibial tuberosity

Anterior intercondylar area

Medial condyle

Lateral condyle

Tibial spines

Posterior intercondylar area

Figure 3.24 The right tibia and fibula. (A) Anterior aspect. (B) Posterior aspect. (C) Superior aspect of the tibia.

The tibia is the larger of the two bones and is situated medial to the fibula. The proximal end of the tibia consists of two large condyles, the lateral and medial condyles, which are fused together side by side (see **Figure 3.24C**). The tibial condyles articulate with the femoral condyles to form the tibiofemoral joint (knee joint). The articular surfaces of the tibial condyles are oval in outline and almost flat. The lateral surface is usually slightly convex and slightly smaller than the medial surface. The medial surface may be slightly convex. The surfaces occupy the same plane more or less horizontally in the anatomical position. This orientation of the condylar surface gives rise to the term tibial table, sometimes used to describe the proximal end of the tibia.

The middle two-thirds of the shaft is teardrop shaped in cross section; the posterior aspect is rounded, whereas the anterior aspect consists of two fairly flat areas, anterior lateral and anterior medial, which converge anteriorly to form a distinct ridge called the anterior crest. The anterior crest can easily be felt beneath the skin as a ridge running down the bone. The anterior medial surface of the tibia, covered only by skin, is usually referred to as the shin. Above and below the anterior crest the shaft broadens out towards the proximal and distal ends of the bone. Above the upper end of the anterior crest, on the anterior aspect of the shaft, is a fairly large process called the tibial tuberosity.

On the medial side of the distal end of the tibia there is a downward projection called the medial malleolus. The lateral aspect of the medial malleolus articulates with the medial aspect of the talus to form the medial part of the ankle joint (see **Figure 3.24**). The remainder of the distal end of the tibia is dominated by a large biconcave condylar surface called the trochlea surface of the tibia. The trochlea surface articulates with the superior aspect of the talus to form the main part of the ankle joint. The trochlea surface of the tibia and the articular surface of the medial malleolus are continuous with each other. The tibia is almost completely responsible for transmitting loads from the upper leg to the foot and vice versa.

The fibula is a thin, relatively weak bone that is only marginally involved in load transmission between the upper leg and foot. The main functions of the fibula are to provide lateral support to the ankle joint and to provide additional area for the attachment of muscles that move the ankle and foot. The proximal end of the fibula is called the head. The medial two-thirds of the superior aspect of the head articulates with the posterior inferior lateral aspect of the lateral tibial condyle to form the proximal tibiofibular joint. The shaft of the fibula is characterized by four longitudinal ridges that give rise to four faces of varying width and length along the shaft. The distal end of the fibula is called the lateral malleolus.

The medial aspect of the lateral malleolus articulates with the lateral aspect of the talus to form the lateral part of the ankle joint. The medial part of the shaft of the fibula immediately above the articular surface of the lateral malleolus articulates with the lateral part of the distal end of the tibia to form the distal tibiofibular joint.

Foot

The foot consists of 7 tarsals, 5 metatarsals and 14 phalanges, often grouped into rearfoot (talus and calcaneus), midfoot (navicular, cuboid, cuneiforms) and forefoot (metatarsals and phalanges) (Briggs 2005) (Figure 3.25). When articulated, the tarsals form the tarsus. The tarsus corresponds to the carpus in the upper limb, but the tarsals are all much larger than the carpals. Whereas the carpus is not usually considered to be part of the hand, the tarsus forms the posterior half of the foot. The foot articulates with the lower leg at the ankle joint, i.e. the joint between the tibia, fibula and talus.

The talus, the second largest tarsal, has a convex pulley-shaped articular surface on its superior aspect called the trochlea surface of the talus (Figure 3.25C) that articulates with the trochlea surface of the tibia. The trochlea surface of the talus is continuous with articular surfaces on its lateral and medial aspects, which articulate with the lateral malleolus and medial malleolus, respectively.

The inferior aspect of the talus articulates with the anterior half of the superior aspect of the calcaneus by means of two or, in some cases, three articular facets, which together constitute the subtalar joint (talocalcaneal joint). The anterior aspect of the talus articulates with the posterior aspect of the navicular, on the medial aspect of the foot, to form the talonavicular joint.

The calcaneus, the largest tarsal, is often referred to as the heel bone. The anterior aspect of the calcaneus articulates with the posterior aspect of the cuboid, on the lateral aspect of the foot, to form the calcaneocuboid joint. The calcaneocuboid and talonavicular joints are continuous with each other and constitute the midtarsal joint, also referred to as the transverse tarsal joint (see Figure 3.25). The anterior aspect of the navicular articulates with the posterior aspects of the three cuneiforms (medial, middle, lateral), which lie side by side and articulate with each other. The posterior two-thirds of the lateral aspect of the lateral cuneiform articulates with the medial surface of the cuboid. The anterior aspect of the cuneiforms articulates with the bases of the first, second and third metatarsals. The anterior aspect of the cuboid articulates with the bases of the fourth and fifth metatarsals. The joints between the four anterior tarsals and the metatarsals are referred to as the tarsometatarsal joints.

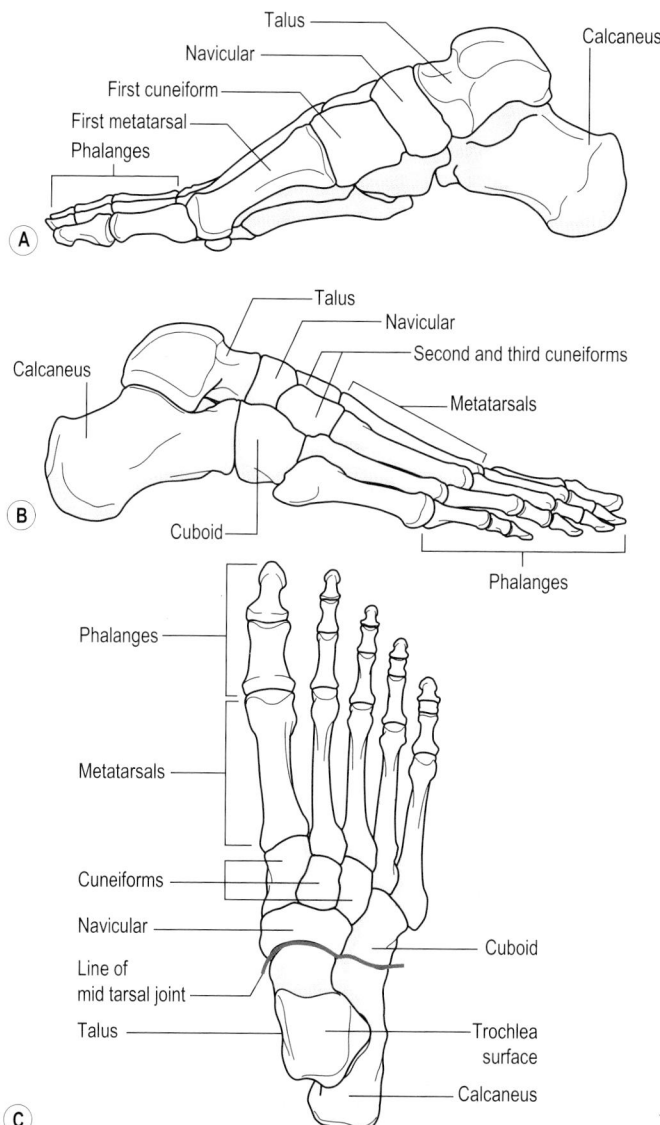

Figure 3.25 The right foot. (A) Medial aspect. (B) Lateral aspect. (C) Superior aspect.

The lateral four metatarsals are similar in length, but tend to increase in girth from the second through to the fifth. In comparison, the first metatarsal is shorter, but has a greater girth than the other four. The metatarsals are collectively referred to as the metatarsus. The heads of the metatarsals articulate with the proximal phalanges of the toes to form the metatarsophalangeal joints.

The distribution of phalanges in the foot is similar to that in the hand, two in the great toe (big toe) and three in each of the other toes. As in the hand, the phalanges of the toes become progressively shorter from proximal to distal. In comparison to the corresponding phalanges of the thumb, the phalanges of the great toe are slightly longer and have a much greater girth. However, the phalanges of the other four toes are much shorter and, in general, smaller in girth than the corresponding phalanges in the hand. The interphalangeal and metatarsophalangeal joints are similar in structure to their counterparts in the hand.

In addition to the tarsals, metatarsals and phalanges, a number of small accessory bones and sesamoid bones occur during fetal life (Williams et al 1995). There are normally about 10 irregular-shaped accessory bones distributed around the tarsus and most of these bones fuse with one of the tarsal bones prior to skeletal maturity. There are normally about 12 sesamoid bones. Like the patella, each sesamoid bone in the foot is partially embedded in a tendon or ligament, with the free surface of the bone forming a synovial joint (sliding between the articular surfaces; see Chapter 5) with a bone over which the tendon or ligament slides during normal function.

Review questions

1. Describe the three main mechanical functions of the skeleton.
2. Describe the three main reference planes and define the spatial terminology associated with the planes.
3. List the bones of the axial skeleton.
4. Describe the primary and secondary curves of the vertebral column.
5. Describe the components of a typical vertebra.
6. Describe the difference between true ribs and false ribs.
7. List the bones of the upper limb.
8. Describe the shoulder girdle.
9. List the bones of the lower limb.
10. Describe the differences between the male pelvis and the female pelvis.

References

Briggs PJ (2005) The structure and function of the foot in relation to injury. Current Orthopaedics 19:85–93.

Watkins J (1999) Structure and function of the musculoskeletal system. Human Kinetics, Champaign, IL.

Williams PL, Bannister LH, Berry MM, et al, eds (1995) Gray's anatomy. Longman, Edinburgh.

Connective tissues

In muscle tissue, nerve tissue and epithelial tissue, the cells are closely joined together by connective tissue in much the same way that concrete joins the bricks in a wall. Consequently, muscle tissue, nerve tissue and epithelial tissue are dominated by the actual tissue cells, with a relatively small amount of connective tissue between the cells. Connective tissue consists of relatively few connective tissue cells and the tissue is dominated by a relatively large amount of non-cellular material called matrix that is produced by the connective tissue cells. It is the matrix of the connective tissue that joins together the cells in muscle tissue, nerve tissue and epithelial tissue and that joins together tissues into organs and organs into systems. The function of a connective tissue is determined by the physical properties of its matrix. The purpose of this chapter is to describe the structure and functions of the various connective tissues.

Functions of connective tissues

Connective tissue, in one form or another, is continuous throughout the body; the composition of the matrix of connective tissue changes from one part of an organ or system to another depending upon function. Connective tissues have two main functions: mechanical support and intercellular exchange.

Mechanical support

Most connective tissues transmit forces to carry out a wide range of mechanical functions. These functions, largely concerned with providing strength or elasticity, include the following:

- Joining together the cells of the body in the various tissues, organs and systems.
- Supporting and holding in place the various organs.
- Linking bones together at joints.
- Providing flexible links between bones in certain joints.
- Providing smooth articulating surfaces between bones in certain joints.
- Providing stability and shock absorption in joints.
- Transmitting muscle forces.

Intercellular exchange

In multicellular organisms, the cells rely on the circulating body fluids such as blood to supply them with nutrients, oxygen and other substances, and to carry away waste products such as carbon dioxide. This involves exchange of nutrients, gases and other substances between the vessels of the circulating body fluids and cells adjacent to the vessels, and between cells adjacent to each other. Intercellular exchange ensures that all cells can be supplied with nutrients, gases and other substances, and can excrete waste products, even if the cells do not receive a direct supply of the circulating body fluids.

Key Concepts

Connective tissues have two main functions: mechanical support and intercellular exchange

Classification of connective tissues

Connective tissues are classified according to level of specialization into ordinary and special connective tissues (Williams et al 1995). Ordinary connective tissues are widely distributed throughout the body; at the tissue level they provide mechanical support and intercellular exchange and at the organ and system level they provide mechanical support. There are two special connective tissues, cartilage and bone. The function of cartilage is to transmit loads across joints efficiently and to allow movement between bones at certain joints. The functions of bone are described in Chapter 3 (The skeleton).

Ordinary connective tissues

The matrix of ordinary connective tissues consists of three basic components: elastin fibres, collagen fibres and ground substance. The main difference in structure of the various types of ordinary connective tissue is in the proportions of these basic components in the matrix.

Elastin and collagen fibres

Elastin fibres and collagen fibres are proteins. A protein molecule consists of a long chain of amino acids. In elastin, the molecules adopt different shapes and are arranged randomly in terms of orientation and attachment to one another (Alexander 1975) (**Figure 4.1A**). When elastin

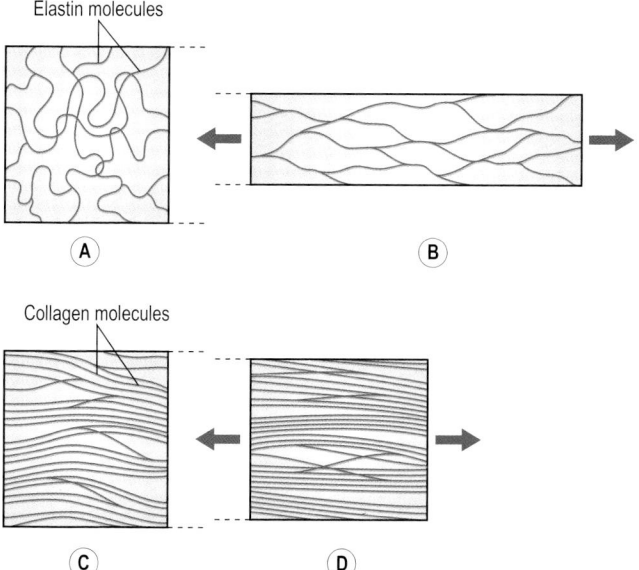

Figure 4.1 (A) Arrangement of molecules in elastin (unloaded). (B) Effect of stretching on elastin molecules. (C) Arrangement of molecules in collagen (unloaded). (D) Effect of stretching on collagen molecules.

is stretched, the molecules do not stretch, but tend to straighten in the direction of stretching (see **Figure 4.1B**). The molecules resist straightening, i.e. they experience tension; the greater the stretch, the greater the tension. When the stretching load is removed, the elastin molecules restore their original orientation and shape. Elastin is, therefore, elastic, hence its name.

An elastin fibril is formed by a number of elastin molecules and an elastin fibre consists of a number of fibrils grouped together. An elastin fibre is similar in shape, strength and elasticity to a long, thin rubber band. Elastin fibres can be stretched by about 200% of their resting length before breaking (Nordin & Frankel 1989). They have a yellowish appearance and are often referred to as yellow elastic fibres or yellow fibres.

In contrast to elastin, collagen molecules are arranged in a more regular manner; they tend to run in the same overall direction and for the most part are aligned parallel to each other (Alexander 1975) (see **Figure 4.1C**). Like elastin molecules, collagen molecules are attached to each other at various points. When stretched in the direction of their main orientation, collagen molecules quickly straighten so that the amount of extension is limited (see **Figure 4.1D**). Like elastin molecules, collagen molecules experience tension when stretched; the greater the stretch, the greater the tension. Like elastin molecules, collagen molecules are also elastic and, as such, return to their resting orientation when the stretching load is removed. Each group of closely aligned parallel molecules constitutes an individual collagen fibril, and a collagen fibre consists of a number of fibrils grouped together. A collagen fibre is similar in shape, strength and elasticity to a shoelace; it is virtually inextensible and, in relation to elastin, it is extremely strong. Collagen fibres break after being stretched by approximately 10% of their rest length (Nordin & Frankel 1989). Collagen fibres are white and are often referred to as white collagen fibres or white fibres.

Ground substance

Ground substance forms the non-fibrous part of the matrix. It is a viscous gel consisting mainly of large carbohydrate molecules (molecules consisting of carbon, hydrogen and oxygen) and carbohydrate–protein molecular complexes (molecules consisting of carbon, hydrogen, oxygen and nitrogen) suspended in a relatively large volume of water (Alexander 1975; Williams et al 1995). The actual volume of water is determined by the number and type of carbohydrate and carbohydrate–protein substances. Many of these substances have an affinity for water and, as such, determine not only the volume of water in the ground substance, but also the viscosity of the ground substance. Viscosity refers to the resistance of a fluid to flowing (how quickly it changes shape in response to external forces) and its stickiness (how strongly it adheres to adjacent structures). For example, oil is more viscous than water.

In contrast to elastin and collagen fibres, whose sole function is to provide mechanical support, ground substance is responsible not only for facilitating intercellular exchange, but also for providing some mechanical support. The glue-like viscosity of ground substance enables it to join cells together within the other main tissues (muscle, nerve and epithelia).

Ground substance in ordinary connective tissues is sometimes referred to as tissue fluid or extracellular fluid. In addition, it is also referred to as amorphous ground substance (amorphous = without definite structure) as it appears, even under a microscope, as a featureless fluid.

Key Concepts

The matrix of ordinary connective tissues consists of elastin fibres, collagen fibres and ground substance. Elastin fibres provide elasticity and collagen fibres provide strength with a small amount of elasticity. Ground substance facilitates intercellular exchange and helps to join cells within the other main tissues

---

Ordinary connective tissue cells

The number and type of cells found in ordinary connective tissues varies according to the type of connective tissue and the state of health of the individual (Williams et al 1995). When present, the various types of cell are found suspended in ground substance or, in some cases, attached to the collagen fibres. In general, there are six main types of cell found in ordinary connective tissues:

- *Fibroblasts*: these are usually the most numerous type of cell in a connective tissue and are responsible for producing the matrix.
- *Macrophages*: these are responsible for engulfing and digesting bacteria and other foreign bodies. They also dispose of dead cellular material that occurs as a result of injury or as cells become old and die.
- *Plasma cells*: these occur in large numbers in response to infection. They produce antibodies that inactivate and, with the macrophages, destroy harmful bacteria and other substances.

- *White blood cells*: the number and type of white blood cells increase in response to infection. They work with the plasma cells and macrophages to identify and destroy harmful bacteria and other substances.
- *Mast cells*: these are widespread throughout ordinary connective tissues and are responsible for producing heparin, which prevents the blood plasma from clotting inside blood vessels.
- *Fat cells*: these have a variety of functions and occur in large numbers in one particular type of ordinary connective tissue (adipose tissue).

The proportion of elastin fibres, collagen fibres, ground substance, and the number and type of cells within any particular ordinary connective tissue determines the function of the connective tissue. Collagen fibres predominate where great strength is required, whereas elastin fibres predominate where considerable elasticity is needed. Similarly, ground substance tends to predominate where intercellular exchange is of major importance. Under normal circumstances a wide variety of cells are present within ordinary connective tissues. In response to infection, there is an increase in the number of cells responsible for identifying and destroying harmful bacteria.

Irregular ordinary connective tissues

Ordinary connective tissues are classified into irregular and regular tissues according to the arrangement of the fibrous content of the matrix. In irregular tissues, the fibres tend to run in all directions throughout the tissue with no set pattern. In contrast, the fibres in regular tissues tend to be orientated in the same overall direction. There are four main types of irregular ordinary connective tissue (loose, adipose, irregular collagenous, irregular elastic) and two main types of regular ordinary connective tissue (regular collagenous, regular elastic). **Figure 4.2** summarizes the six main types of ordinary connective tissue in relation to dominant feature and fibre arrangement.

Loose connective tissue

Loose connective tissue is the most widely distributed of all the connective tissues. It is the predominant type of connective tissue that joins the cells in the other main tissues (muscle, nerve and epithelia) and that joins tissues into organs. It consists of a loose irregular network of elastin fibres and collagen fibres suspended within a relatively large amount of ground substance (**Figure 4.3**). The large amount of amorphous ground substance gives the impression of a lot of space between the fibres and cells of loose connective tissue. For this reason, loose connective tissue is also referred to as areolar tissue (areola = a small open area).

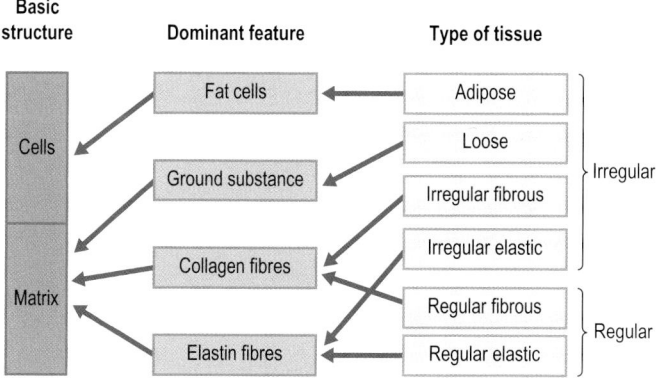

Figure 4.2 Ordinary connective tissues.

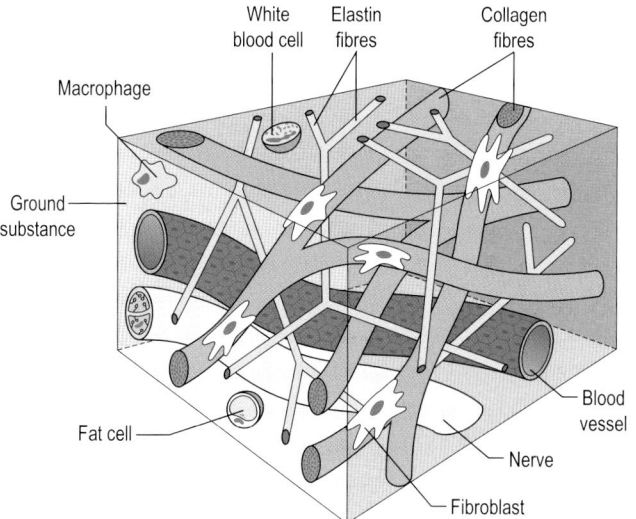

Figure 4.3 Loose connective tissue.

In association with ground substance, the loose network of elastin fibres and collagen fibres, both of which branch freely, provides moderate amounts of elasticity and strength in tissues and organs. In addition, the semi-liquid form of loose connective tissue provides a supporting

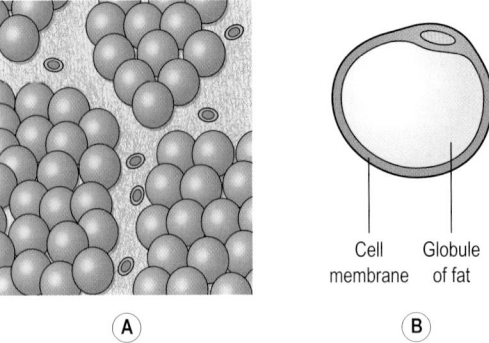

Cell | Globule
membrane | of fat

(A)   (B)

Figure 4.4 Adipose connective tissue. (A) Groups of fat cells held together by loose connective tissue. (B) Cross section of a single fat cell.

framework for nerves, blood vessels and lymph vessels. The large amount of ground substance reflects the importance of loose connective tissue in the facilitation of intercellular exchange.

Adipose connective tissue

Adipose tissue has a loose network of elastin and collagen fibres similar to loose connective tissue. However, in contrast to loose connective tissue, there is little ground substance and a large number of closely packed fat cells. Each fat cell consists of a thin cell membrane surrounding a relatively large globule of fat (**Figure 4.4**). Adipose tissue is widely distributed around the body, particularly in the following locations (McArdle et al 1996):

1. In bone marrow.
2. In association with the various layers of loose connective tissue within certain organs, especially skeletal muscles.
3. As padding around certain organs and joints.
4. As a continuous layer beneath the skin. Skin is sometimes referred to as cutaneous tissue and the layer of fat as the subcutaneous fat layer.

Adipose tissue is a poor conductor of heat and, consequently, the subcutaneous fat layer acts as an insulator, reducing the loss of body heat through the skin. Adipose tissue is moderately strong due to its

Collagen fibres   Fibroblast

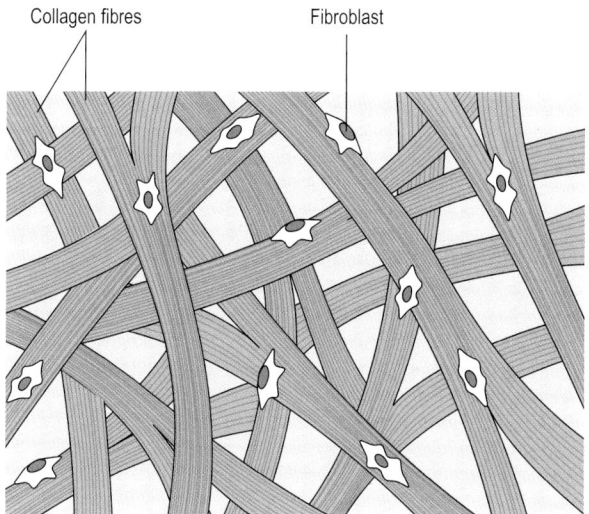

Figure 4.5 Irregular collagenous connective tissue.

collagen fibre content and considerably elastic due to its elastin fibre content and the large number of elastic fat cells. Consequently, adipose tissue is well suited to provide mechanical support and protection (cushioning) in the form of padding around and between certain organs such as the heart, lungs, liver, spleen, kidneys and intestines. Adipose tissue also acts as padding around joints such as the knee and over certain bones such as the heel bone. In addition to its heat insulation and mechanical functions, adipose tissue is the body's main food store. Adipose tissue provides approximately twice as much energy per gram as any other tissue in the body (McArdle et al 1996).

Irregular collagenous connective tissue

The matrix of irregular collagenous connective tissue is dominated by a dense, irregular network of bundles of collagen fibres with few elastin fibres and little ground substance (**Figure 4.5**). The collagen bundles and their irregular arrangement enable the tissue to resist being stretched in any direction. However, though it is strong, the tissue has a certain amount of elasticity due to the wavy orientation of the collagen bundles. When stretched in a particular direction, the collagen bundles tend to straighten in the direction of stretching. Irregular collagenous connective tissue is most frequently found as a tough cover around certain organs, where it provides mechanical support and protection. For example, it is found as:

- A sheath around skeletal muscles (epimysium) and spinal nerves (epineurium).
- A capsule or envelope around certain organs such as the kidneys, liver and spleen that holds the organs in place.
- The perichondrium of cartilage (discussed later in this chapter).
- The periosteum of bones (discussed later in this chapter).

Irregular elastic connective tissue

The matrix of irregular elastic connective tissue is dominated by a dense, irregular network of bundles of elastic fibres with few collagen fibres and a moderate amount of ground substance (**Figure 4.6**). In comparison to irregular collagenous connective tissue, irregular elastic connective tissue is not as strong, but much more elastic. It is found where moderate amounts of strength and elasticity are required in more than one direction

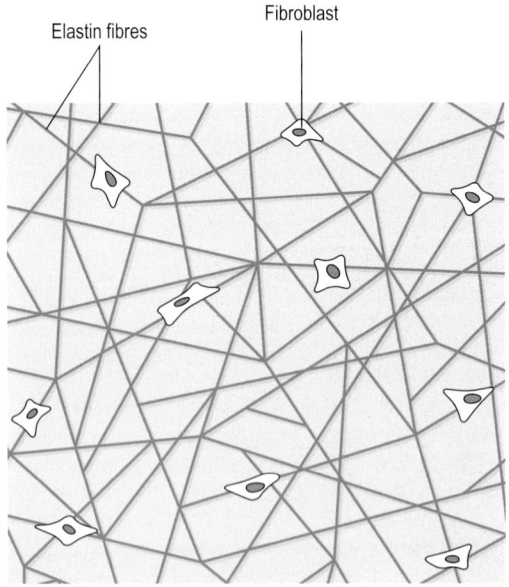

Elastin fibres   Fibroblast

Figure 4.6 Irregular elastic connective tissue.

as, for example, in the walls of arteries and the larger arterioles, the trachea (windpipe) and bronchial tubes. There are few cells in irregular collagenous and irregular elastic connective tissues. The cells that are present are mainly fibroblasts.

Regular ordinary connective tissues

There are two main types of regular ordinary connective tissue: regular collagenous and regular elastic.

Regular collagenous connective tissue

Regular collagenous connective tissue consists almost entirely of bundles of collagen fibres arranged parallel to each other. Usually, there are few elastic fibres and little ground substance. The only cells present are fibroblasts arranged in columns between the collagen bundles (Figure 4.7). The collagen bundles are gathered together in the form of thick cords, bands or sheets of various widths. In the unloaded state the collagen bundles have a slightly wavy orientation. When stretched, the bundles quickly straighten,

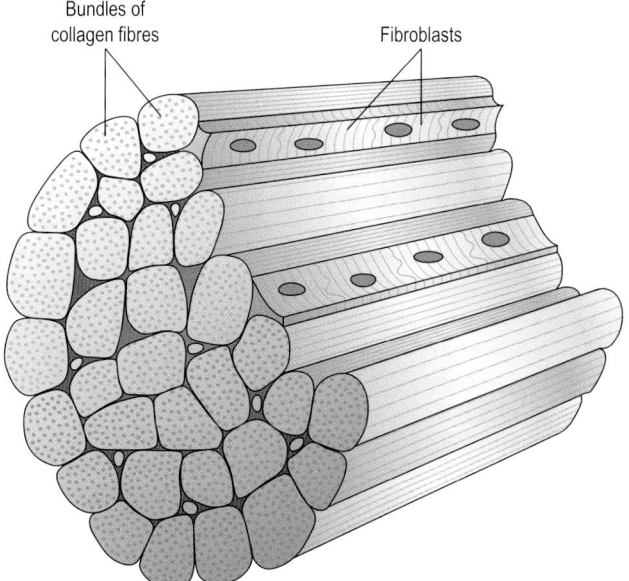

Bundles of
collagen fibres

Fibroblasts

Figure 4.7 Regular collagenous connective tissue: part of a tendon.

and the tissue becomes taut. Regular collagenous connective tissue is extremely strong and virtually inextensible. It has three main forms:

1. Tendons and aponeuroses: mechanical links between skeletal muscle and bone.
2. Ligaments and joint capsules: mechanical links between bones at joints.
3. Retinacula: mechanical restraints on tendons that increase the mechanical efficiency of the muscle–tendon units.

Tendons and aponeuroses Skeletal muscles are attached to the skeleton by regular collagenous connective tissue in the form of tendons and aponeuroses (Figure 4.8).

Ligaments and joint capsules Skeletal muscles provide active (contractile) links between bones. Ligaments (Latin: *ligare* = to bind) and joint capsules provide passive (non-contractile) links between bones. In association

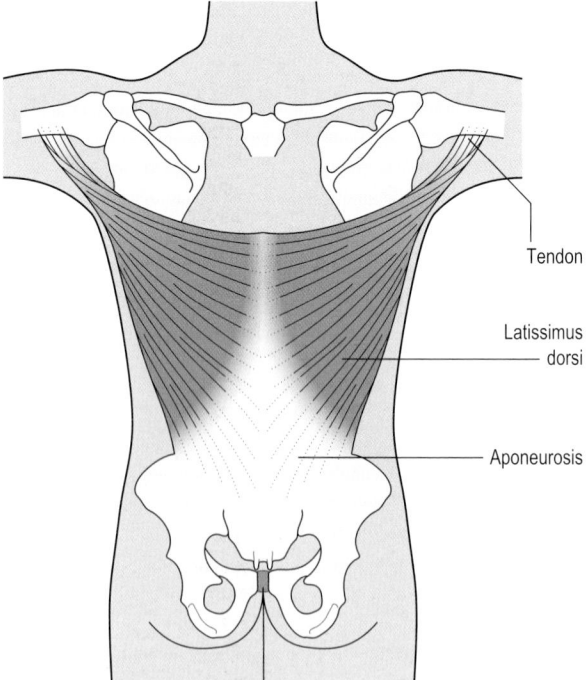

Figure 4.8 The latissimus dorsi muscles.

with skeletal muscles, ligaments and joint capsules bring about normal movements in joints. Chapter 5 (The articular system) describes the different types of joints in the body. At this point it is sufficient to appreciate that each synovial (freely moveable) joint, i.e. a joint involving sliding or rolling between the free surfaces of the ends of bones, as in the shoulder and hip, is enclosed within its own joint capsule (Figures 4.9 and 4.10).

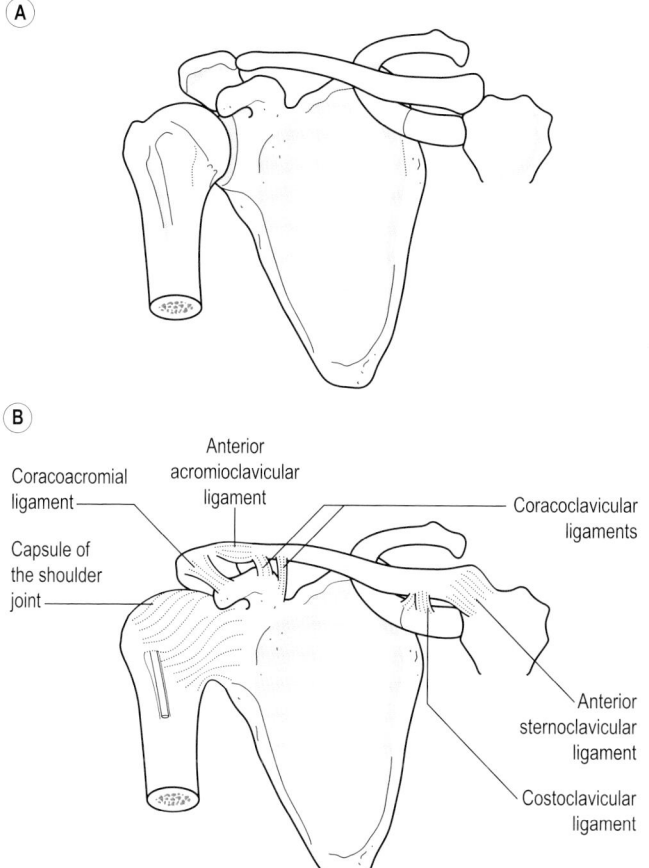

Figure 4.9 Ligaments of the shoulder girdle. (A) Anterior aspect of the right shoulder girdle and right shoulder joint. (B) Shoulder joint capsule and ligaments supporting the shoulder girdle.

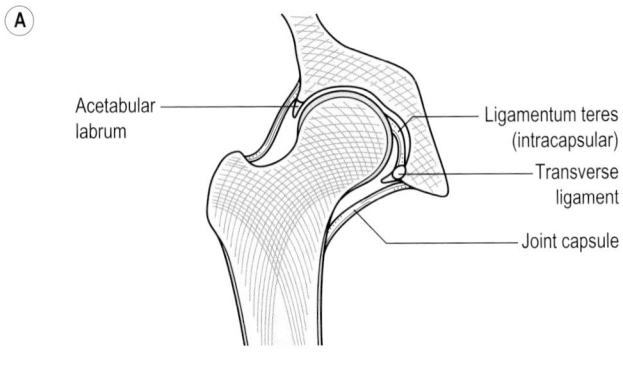

(A)

Acetabular labrum

Ligamentum teres (intracapsular)

Transverse ligament

Joint capsule

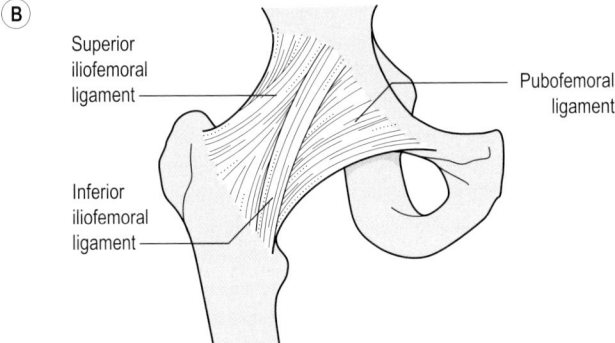

(B)

Superior iliofemoral ligament

Pubofemoral ligament

Inferior iliofemoral ligament

Figure 4.10 Ligaments of the hip joint. (A) Coronal section through the right hip joint showing the joint capsule and ligamentum teres. (B) Anterior aspect of the right hip joint showing the joint capsule and anterior capsular ligaments.

The joint capsule encloses a space, usually quite small, called the joint cavity. A joint capsule is composed of two or more layers of regular collagenous connective tissue forming a sleeve around the joint, rather like a piece of rubber tubing joining two glass rods together. Whereas the collagen bundles in each layer are parallel to each other, the bundles in adjacent layers run in different directions. This arrangement enables the capsule to strongly resist stretching in a number of different directions and, therefore, helps to maintain joint integrity (Figure 4.11).

In all synovial joints, the joint capsule is supported by a number of ligaments. These ligaments may be capsular or non-capsular. A capsular

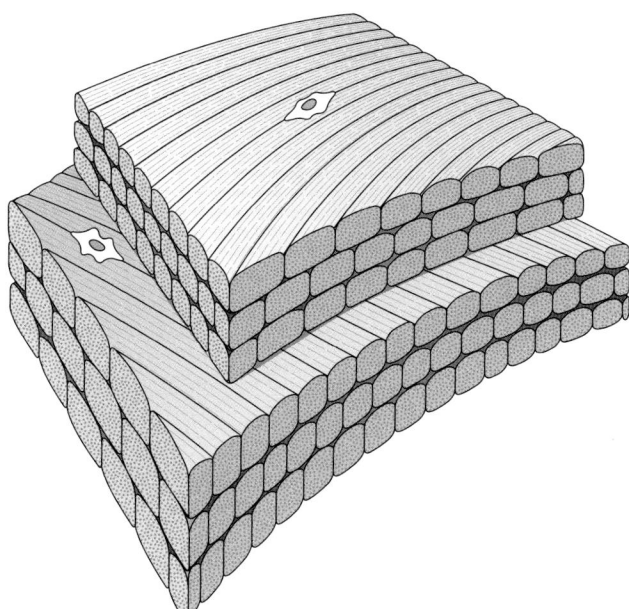

Figure 4.11 Three-dimensional section through a two-layered joint capsule.

ligament is a distinct thickening in part of the joint capsule that provides additional strength in one direction. For example, the superior iliofemoral ligament, inferior iliofemoral ligament and pubofemoral ligament are capsular ligaments that strengthen the anterior aspect of the capsule of the hip joint (see Figure 4.10B). A non-capsular ligament is a distinct band separate from the joint capsule or only partially attached to it. Non-capsular ligaments may be extracapsular (outside the joint cavity) or intracapsular (inside the joint cavity). For example, the ligamentum teres of the hip joint is intracapsular (see Figure 4.10A), but the lateral ligament, medial ligament and cruciate ligaments of the knee joint are all extracapsular (Figure 4.12). Non-capsular ligaments usually consist of a single layer of tissue, but broad ligaments may consist of two or more layers, similar to a joint capsule.

In addition to the non-capsular ligaments associated with synovial joints, there are other ligaments similar in structure to non-capsular ligaments that help stabilize other parts of the skeleton; for example, the

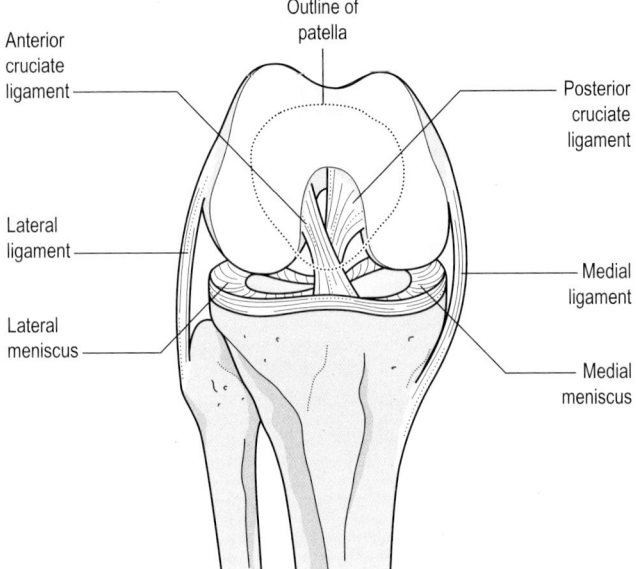

Figure 4.12 Extracapsular ligaments of the knee joint. Anterior aspect of the right knee, flexed at 90°, with patella removed to show the cruciate ligaments and femoral condyles slightly raised to show the menisci. The cruciate ligaments are located at the centre of the joint, but lie outside the joint cavity.

ligaments between the clavicle and scapula, and between the clavicle and first rib (see Figure 4.9B).

Retinacula A retinaculum (Latin: *retinere* = to retain) is a fairly broad, single-layered sheet of regular collagenous connective tissue that restrains the tendons of some muscles so that the tendons operate close to the joint(s) that they cross. There are basically two forms of retinacula:

1. In the form of a guy rope that restricts the side-to-side movement of a tendon. For example, there are two retinacula, one on each side of the knee joint, that restrain the patella and, therefore, the quadriceps tendon (Figure 4.13). These retinacula help to maintain normal movement between the patella and the femur during flexion and extension of the knee.

2. In the form of a pulley that prevents one or more tendons from springing away from a joint when the muscles contract. For example, the

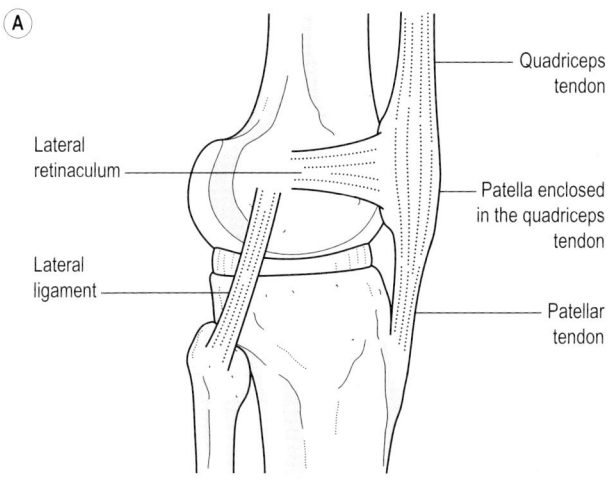

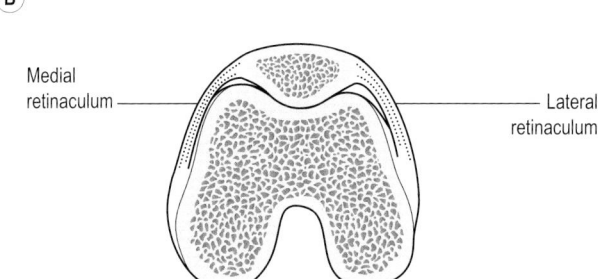

Figure 4.13 Retinacula of the patellofemoral joint. (A) Lateral aspect of the right knee joint. (B) Superior aspect of a transverse section through the patellofemoral joint of the right knee joint.

retinacula at the wrist and ankle hold down the tendons of the muscles that move these joints (**Figure 4.14**). The effect of this form of retinaculum is to increase considerably the mechanical efficiency of the associated muscles (see length–tension relationship of skeletal muscles in Chapter 6).

Regular elastic connective tissue

Regular elastic connective tissue consists largely of elastin fibres arranged parallel to each other. The proportion of collagen fibres and ground

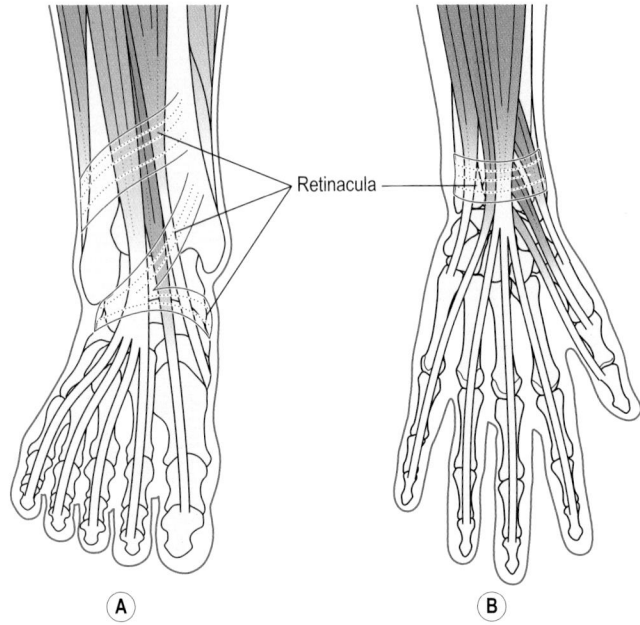

Figure 4.14 Retinacula at the ankle (A) and wrist (B).

substance is usually fairly small. However, the proportion of collagen fibres and ground substance in regular elastic connective tissue is usually greater than the proportion of elastic fibres and ground substance in regular collagenous connective tissue (Akeson et al 1985). Regular elastic connective tissue is found where moderate amounts of strength and elasticity are required mainly in a single direction. Most ligaments consist of regular collagenous connective tissue, but a few consist of regular elastic connective tissue. Two of these so-called elastic ligaments (ligamentum nuchae, ligamentum flavum) help to stabilize the vertebral column and to allow a certain amount of movement between the vertebrae.

Fibrous tissue, elastic tissue and fascia

Four of the six main types of ordinary connective tissue are dominated by collagen or elastic fibres: irregular collagenous, irregular elastic, regular collagenous and regular elastic. Whereas all of these tissues could be described as fibrous, the term fibrous tissue normally refers only to regular or irregular collagenous connective tissue. Elastic tissue normally refers only to regular or irregular elastic connective tissue.

The term fascia (Latin: *fascia* = band) refers to any type of ordinary connective tissue in the form of a sheet. In this sense, all aponeuroses are fascia. However, fascia most often refers to superficial fascia and deep fascia. Superficial fascia refers to the continuous layer of loose connective tissue that connects the skin to underlying muscle or bone. This layer of loose connective tissue is closely associated with the subcutaneous layer of fat referred to earlier. Deep fascia describes the sheets of irregular collagenous connective tissue that form sheaths around muscles and groups of muscles, separating them into functional units.

Key *Concepts*

Ordinary connective tissues differ from each other in the proportions of elastin, collagen and ground substance, and in the arrangement of the fibrous content. There are four main types of irregular ordinary connective tissue – loose, adipose, irregular collagenous and irregular elastic – and two main types of regular ordinary connective tissues – regular collagenous and regular elastic

Cartilage

Like all connective tissues, cartilage is a composite material, i.e. a material that is stronger than any of the separate substances from which it is made (Alexander 1968). Fibreglass and the rubber used in the manufacture of tyres are examples of manmade composite materials. The matrix of cartilage is similar to that of fibrous and elastic ordinary connective tissues in that it consists mainly of collagen and elastin fibres embedded in ground substance. However, relative to ordinary connective tissues, the ground substance of cartilage is specialized to produce a material that is capable of resisting all forms of loading, not just tension (Caplan 1984). The ground substance of cartilage consists of huge carbohydrate–protein molecular complexes called proteoglycans that have a high affinity for water; each proteoglycan complex is capable of attracting to itself a volume of water that is many times its own weight. Consequently, under normal circumstances, water is the chief constituent of cartilage. The proteoglycans and water produce a highly viscous gel usually referred to as proteoglycan gel. In combination with collagen and elastin, the proteoglycan gel forms a tough rubbery material capable of strongly resisting all forms of loading.

The only cells found in cartilage are cartilage cells, called chondrocytes, which produce the cartilage matrix. The chondrocytes lie in fluid-filled spaces called lacunae distributed throughout the matrix (Figure 4.15). The cells are arranged singly (parent cells) and in groups of two to five cells that originate from a single parent cell. As the cells become mature they separate from their parent groups and start to produce new groups. Whereas collagen and elastin fibres dominate the matrix of all three main types of cartilage, the region around each lacuna is usually free of fibres. This distinct region, called the capsule of the lacuna, consists of proteoglycan gel that is denser than in other parts of the matrix.

Mature cartilage contains no blood vessels or nerves (Nordin & Frankel 1989).This reflects the mechanical functions of cartilage, in the sense that blood vessels and nerves would be destroyed by the deformation of cartilage in response to loading. In the absence of a direct blood supply, the chondrocytes depend on intercellular exchange via blood vessels close to the non-load-bearing surfaces of the cartilage and the proteoglycan gel for their nutrition and excretion. For this reason, and the fact that cartilage is often under considerable load, repair of cartilage is slow and may not take place at all (Caplan 1984).

When cartilage is loaded (subjected to tension, compression, shear, or any combination of these loads), water is forced out of the cartilage and the cartilage deforms. The rate and extent of deformation depends

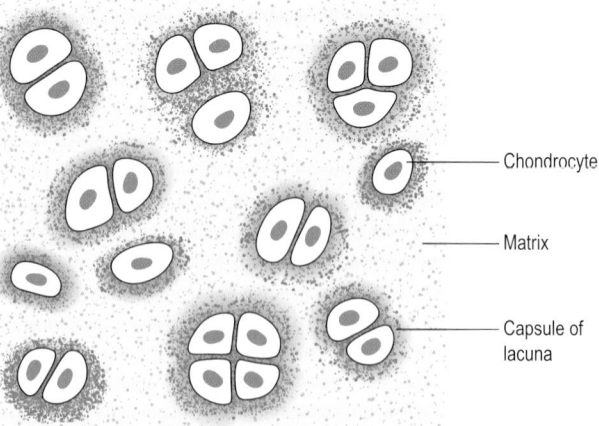

Chondrocyte

Matrix

Capsule of lacuna

Figure 4.15 Typical structure of cartilage.

on the size and duration of the load. When the load is removed the proteoglycan structures restore the original level of water saturation and, consequently, the original size and shape of the cartilage, by absorbing water into the cartilage. The ability of a material to deform gradually in response to a load and, following unloading, to restore gradually its original size and shape is referred to as viscoelasticity. In comparison, elasticity refers to a material's ability to deform immediately in response to a load and, following unloading, to immediately restore its original size and shape, like a rubber band. Cartilage tends to behave viscoelastically in response to prolonged loading, and elastically in response to sudden impact loads.

The physical properties of cartilage (including viscoelasticity, elasticity, tensile strength, compressive strength and shear strength) depend on the proportions of collagen fibres, elastin fibres and proteoglycan gel in the matrix. In general, there are three main types of cartilage: hyaline cartilage, fibrocartilage and elastic cartilage.

Hyaline cartilage

Hyaline cartilage has a pearly bluish-white tinge and under a low-power microscope the matrix appears amorphous and translucent (semitransparent) as in Figure 4.15. Under a high-power microscope the matrix can be seen to consist of a dense network of very fine collagen fibrils and fibres embedded in proteoglycan gel (Williams et al 1995). Most of the skeleton is preformed in hyaline cartilage and prior to maturity the growth and development of many bones is largely determined by the hyaline cartilage content of the bones. The following structures consist of hyaline cartilage throughout life:

- Articular cartilage: the smooth, tough, wear-resistant articulating surfaces of bones in synovial joints.
- The costal cartilages that link the upper 10 pairs of ribs to the sternum and provide the rib cage with flexibility and elasticity.
- Supporting rings within the elastic walls of the trachea (windpipe) and the larger bronchial tubes.
- Part of the supporting framework of the larynx (voice box).
- The external flexible part of the nose that forms the major part of the nostrils.

Fibrocartilage

In fibrocartilage, also referred to as white fibrocartilage, the matrix is dominated by a dense regular network of bundles of collagen fibres arranged parallel to each other in several layers (**Figure 4.16**). The bundles in adjacent layers run in different directions (like the layers in a joint capsule), which produces a strong material with moderate elasticity. Fibrocartilage is found in a number of locations and forms, in particular:

1. Articular discs: complete or incomplete discs interposed between the articular surfaces of some synovial joints including the knee joint, the sternoclavicular joint and the acromioclavicular joint (see Chapter 5). In these joints, the discs improve the congruence (area over which the joint reaction forces are distributed) and stability of the joints. In addition, the discs deform in response to loading and thereby provide shock absorption.

2. Symphysis joints: complete discs that join the bones in symphysis joints (see Chapter 5). These joints include the pubic symphysis and the joints between the bodies of the vertebrae. In these joints, deformation of the disc in response to loading allows movement between the articulated bones and provides shock absorption.

3. Labra in ball and socket joints: extension to the concave articular surface in a ball and socket joint (e.g. shoulder and hip joints) in the

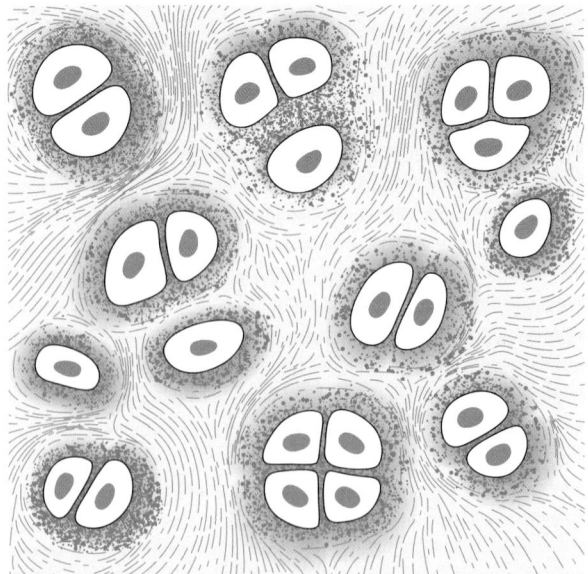

Figure 4.16 Fibrocartilage.

form of a labrum (lip) around the border of the articular surface (see Chapter 5). A labrum deepens the socket, which increases the area of articulation and, therefore, the stability of the joint.

4. Lining of bony grooves: lining of bony grooves and channels that are occupied by tendons, such as the tendons of the tibialis posterior, flexor digitorum longus and flexor hallucis longus in a groove behind the medial malleolus (see **Figure 1.18A**). The grooves act as pulleys that normally increase the mechanical efficiency of the associated muscles.

Elastic cartilage

In elastic cartilage, also referred to as yellow elastic cartilage, the matrix is dominated by a dense network of elastin fibres (**Figure 4.17**). Elastic cartilage provides support with moderate elasticity. It is found mainly in the larynx, the external part of the ear (pinna), and the tube leading from the middle part of the ear to the throat (eustachian or auditory tube).

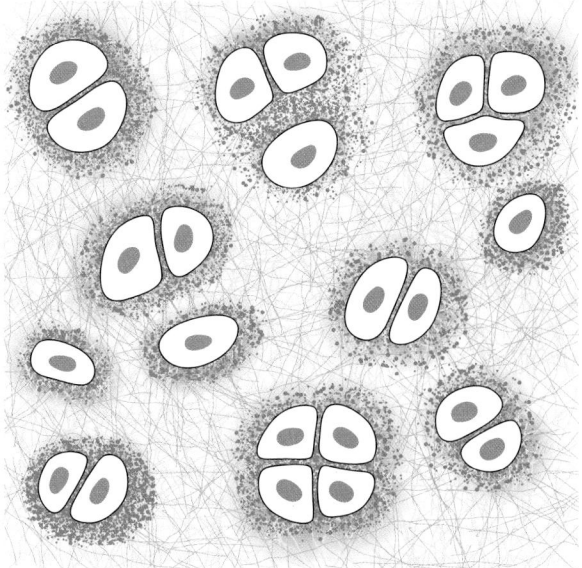

Figure 4.17 Elastic cartilage.

Key Concepts

The physical properties of cartilage depend on the proportions of collagen fibres, elastin fibres and proteoglycan gel in the matrix. There are three main types of cartilage: hyaline cartilage, fibrocartilage and elastic cartilage

Bone

The strongest and least flexible of all the connective tissues is bone. The matrix of bone consists of a dense, layered, regular network of collagen fibres embedded in a hard, solid ground substance called bone salt. Bone salt consists of calcium phosphate and calcium carbonate with smaller amounts of magnesium, sodium and chlorine (Alexander 1975). In mature bone, bone salt makes up about 70% of the total weight of bone, with collagen making up the remaining 30%. Bone salt is denser than collagen such that the bone salt and collagen both occupy about 50% of the total volume. The composite material made of bone salt and collagen produces a hard, tough, fairly stiff structure. Relative to cast iron, bone has the same tensile strength, is only one-third as heavy and is much more elastic (Ascenzi & Bell 1971). The elasticity of bone, although slight relative to cartilage, is nevertheless important in enabling it to absorb sudden impacts without breaking.

Bone growth and development

Most of the embryonic skeleton, which appears around the third week of intrauterine life, is preformed in hyaline cartilage. Those parts of the embryonic skeleton that are not preformed in hyaline cartilage, which include the top of the skull, the clavicles and parts of the mandible, are preformed in a highly vascular form of tissue called fibrous membrane. By the eighth or ninth week of intrauterine life the shapes of the embryonic bones are similar to their eventual adult shape (Williams et al 1995). The developing bones constantly adapt their shapes and structures to withstand the forces (especially muscle forces and joint reaction forces) that act on them. After skeletal maturity (approximately 20 to 25 years of age) the bones experience negligible change in shape, but change in structure continues throughout life (Frost 2004).

Ossification

Ossification or osteogenesis (osteo = bone, genesis = creation) is the process that transforms the embryonic skeleton into bone. Ossification is a feature of normal growth and development and proceeds at different rates in different bones. In each bone, ossification is initiated at a particular location, referred to as the primary centre of ossification. In some bones, including the carpals and tarsals, ossification is completed from the primary centre of ossification, i.e. the volume of ossified bone around the primary centre of ossification gradually increases as the bone matures. In other bones, one or more secondary centres of ossification occur some time after the occurrence of the primary centre of ossification, i.e. ossification in these bones gradually progresses from two or more centres of ossification within the same bone. For example, in each metacarpal and metatarsal there is a primary centre of ossification and a secondary centre of ossification. In all of the large long bones there is a primary centre of ossification in the shaft and a secondary centre of ossification in each epiphysis. At birth, there are approximately 270 regions of the immature skeleton (associated with 270 primary and secondary centres of ossification) that are in a fairly advanced stage of ossification. The 270 regions normally result in 206 bones in the mature skeleton.

The ossification of hyaline cartilage is called intracartilagenous or endochondral ossification (endo = within, chondral = cartilage) and the ossification of fibrous membranes is called intramembranous ossification. Both forms of ossification are similar and produce the same type of bone tissue. The process of endochondral ossification is described with reference to a typical long bone.

Growth in girth

Each embryonic long bone is preformed in hyaline cartilage and covered in a fibrous perichondrium (peri = around, chondrium = cartilage), which contains blood vessels (**Figure 4.18A**). Between the fifth and twelfth weeks of intrauterine life, some fibroblasts in the perichondrium around the middle of the shaft of the cartilage model are transformed into osteoblasts. Osteoblasts are one of three types of bone cell and are responsible for the production of bone. The newly formed osteoblasts invade the hyaline cartilage immediately beneath the perichondrium and start to deposit calcium and other minerals in the matrix. Consequently, the hyaline cartilage is transformed into calcified cartilage. This process of mineralization is called calcification; calcified cartilage represents an intermediate stage in the process of ossification of cartilage into bone. Calcification continues until the calcified cartilage is transformed into bone. Consequently, a bony ring or collar is formed around the middle of the shaft of the otherwise cartilage model (see **Figure 4.18B**).

When the perichondrium starts to produce osteoblasts and, in turn, bone, it is called periosteum. The first site of bone formation, the

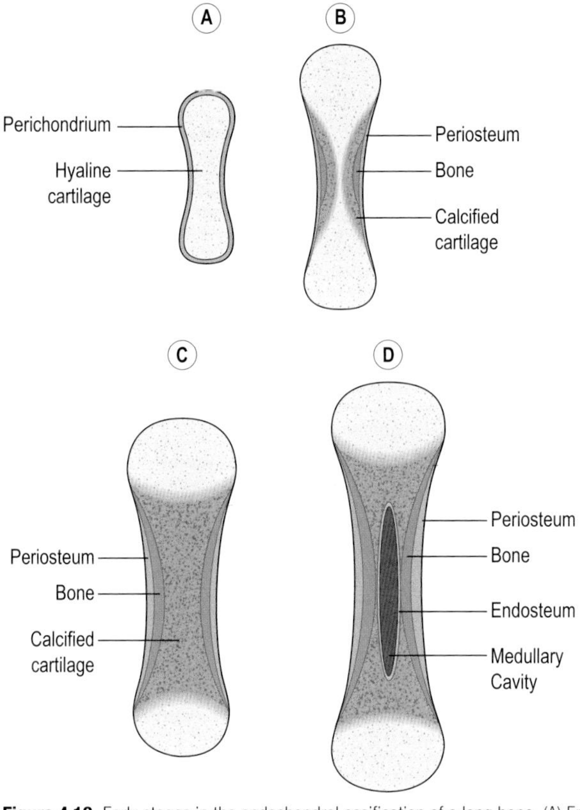

Ⓐ   Ⓑ

Perichondrium ——

Hyaline ——
cartilage

—— Periosteum

—— Bone

—— Calcified
cartilage

Ⓒ   Ⓓ

—— Periosteum

—— Bone

Periosteum ——

Bone ——

Calcified ——
cartilage

—— Endosteum

—— Medullary
Cavity

Figure 4.18 Early stages in the endochondral ossification of a long bone. (A) Embryonic long bone; hyaline cartilage enclosed within a fibrous perichondrium. (B) Establishment of a bony collar around the middle of the shaft. (C) Completion of a bony cylinder running the length of the shaft around the time of birth. (D) Formation of a medullary cavity.

middle of the shaft of the cartilage model, is called the primary centre of ossification. The process of ossification proceeds from the bony collar in two directions: across the shaft from the outside towards the centre and towards the ends of the shaft. By the 36th week, around the time of birth, the bony collar has become a bony cylinder running the length of the shaft, but not progressing into the bulbous ends of the bone (see **Figure 4.18C**). The bony cylinder is thickest at its middle and thinnest at its ends. By this time the remaining hyaline cartilage in the middle of the shaft has been transformed into calcified cartilage. Soon afterwards, a second type of bone cells called osteoclasts invades this central portion of calcified cartilage. Whereas osteoblasts produce new bone, osteoclasts remove bone and calcified cartilage. The osteoclasts start to remove the calcified cartilage in the middle of the shaft thereby creating a space called the medullary cavity (see **Figure 4.18D**). The medullary cavity gradually widens and extends towards both ends of the shaft. Simultaneously, the thickness of bone in the shaft gradually increases.

Eventually all of the calcified cartilage is removed due to the combined effects of ossification across the shaft from the outside towards the centre and osteoclastic activity from the centre outward. By this time, the medullary cavity is occupied by yellow marrow consisting of loose connective tissue containing a large number of blood vessels, fat cells and immature white blood cells (Williams et al 1995). A layer of loose connective tissue containing many osteoblasts and a smaller number of osteoclasts lines the medullary cavity. This layer is called the endosteum.

Long bones are designed, above all, to resist bending. For a given amount of bone tissue, a hollow shaft is stronger in bending than a solid one which would appear to be the main reason why the shafts of long bones are hollow (Alexander 1968). Growth in girth of the shaft of a long bone involves formation of new bone on the outside of the shaft by osteoblasts in the periosteum and the removal of bone from the inside of the shaft by osteoclasts in the endosteum. The type of growth produced by the periosteum, which involves laying down new bone on the surface of older bone rather like the addition of rings in a tree, is called appositional growth. In mature bone, the periosteum consists of irregular collagenous connective tissue. In addition to appositional growth, the periosteum has three other main functions:

1. To provide a protective cover around the shaft of the bone.
2. To allow blood vessels to pass into the bone.
3. To provide attachments for muscles, tendons, ligaments and joint capsules.

Growth in length

Around the time of birth a secondary centre of ossification occurs in the centre of each end of a long bone. These new centres of ossification are responsible for the ossification of the ends of the bones; ossification proceeds from the centre towards the periphery. After the secondary centres of ossification have been established, the only hyaline cartilage remaining from the original cartilage model is the cartilage that covers the bulbous ends of the bone and that which separates the ends of the bone from the shaft (Figure 4.19). These two regions of hyaline cartilage are continuous with each other and remain so until maturity.

Each end of a bone is called an epiphysis and the shaft is called the diaphysis. Part of the cartilage covering each epiphysis forms an articular surface and, as such, is referred to as articular cartilage. The cartilage that separates the epiphyses and diaphysis are called epiphyseal plates (see Figure 4.19). The epiphyseal plates are responsible for growth in length of the bone. During normal growth they remain active until the bone has achieved its mature length.

An epiphyseal plate consists of four layers (Tortora & Anagnostakos 1984) (Figure 4.20). The layer adjacent to the epiphysis is called the reserve or germinal layer; this layer anchors the epiphyseal plate to the bone of the epiphysis. The second layer, called the proliferation layer, is responsible for chondrogenesis (production of new cartilage). The chondrocytes in this layer undergo fairly rapid cell division, and in turn, the cells produce new matrix that results in an increase in the amount of cartilage. Growth in length of the shaft of a bone is due to chondrogenesis in the

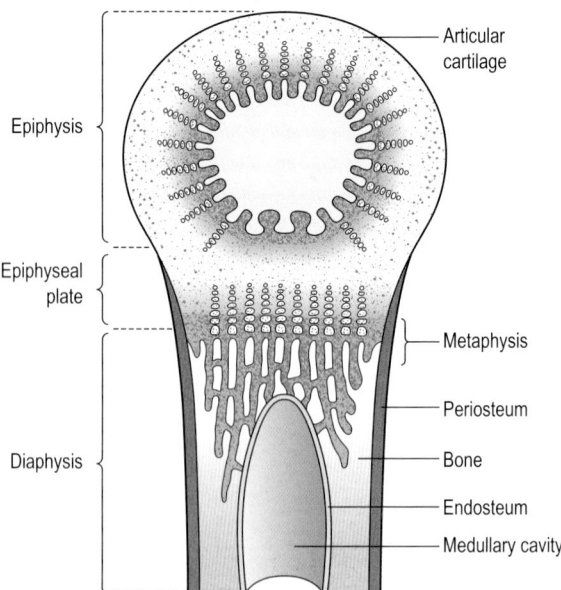

Figure 4.19 Longitudinal section through the epiphysis and part of the diaphysis of a typical immature long bone.

proliferation layers of the epiphyseal plates. This type of growth, in which additional new tissue is produced from within the mass of existing tissue, is called interstitial growth.

The third layer of the epiphyseal plate is called the hypertrophic layer. In this layer, the chondrocytes are arranged in columns and gradually increase in size, with the larger and more mature cells furthest from the epiphysis. The fourth layer of the epiphyseal plate is called the calcified layer. In this layer, the hypertrophied chondrocytes and surrounding matrix are replaced by calcified cartilage. The calcified cartilage interdigitates with the underlying bone forming a relatively strong bond (see **Figures 4.19 and 4.20**) that is able to resist shear loading. As new cartilage is formed in the proliferation layer, the calcified cartilage in contact with the underlying bone is itself gradually transformed into bone. The net result of these processes is that the epiphyseal plates, which remain about the same thickness, gradually move further from the middle of the shaft as the shaft increases in length.

The metaphysis is the region where the epiphysis joins the diaphysis; in a growing bone this corresponds to the calcified layer of the epiphyseal plate together with the interdigitating bone (see **Figure 4.19**). The interface between the hypertrophic and calcified layers is sometimes referred to as the tidemark.

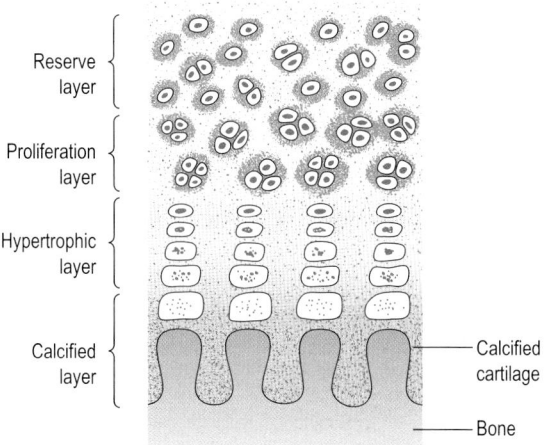

Figure 4.20 A longitudinal section through an epiphyseal plate.

Key *Concepts*

The adult skeleton develops from an embryonic skeleton that pre-forms in hyaline cartilage and fibrous membrane around the third week of intrauterine life. Growth and development of the skeleton involves progressive ossification. Growth in girth of bones occurs by appositional growth; growth in length of bones occurs by interstitial growth

When a long bone has achieved its mature length, longitudinal growth in the epiphyseal plates ceases. Shortly afterwards, the epiphyseal plates are replaced by bone so that the epiphyses are fused with the shaft. In most long bones, one end usually fuses with the shaft before the other end. In the long bones of the arms and legs, fusion of both ends normally takes place between 14 and 20 years of age (Williams et al 1995). In some other bones, such as the innominate bones, fusion takes place usually between 20 and 25 years of age. Consequently, the epiphyseal plates of the various bones are vulnerable to injury for a relatively long period. Injury to an epiphyseal plate may, in severe cases, result in one of two types of bone deformity (Peterson 2001):

1. A complete cessation of growth and premature fusion resulting in, for example, a limb length discrepancy.

2. An asymmetric cessation of growth across an epiphyseal plate resulting in an angular deformity and joint incongruity.

The epiphyseal plates at each end of a long bone usually contribute different amounts to the length of the shaft. For example, the upper and lower epiphyseal plates of the humerus contribute approximately 80% and 20%, respectively, to the total length of the bone. In contrast, the upper and lower epiphyseal plates of the femur contribute approximately 30% and 70%, respectively, to the total length of the bone (Pappas 1983) (Figure 4.21).

The vulnerability of epiphyseal plates to injury is largely due to the plates being the weakest parts of the immature skeleton. For example, ligaments and joint capsules are two to five times stronger than epiphyseal plates (Larson & McMahan 1966). When a ligament supporting a particular joint is inserted into the epiphysis (rather than the diaphysis), a load applied to the joint that tends to stretch the ligament is, in a child, more likely to result in a fracture through the epiphyseal plate than in a tear in the ligament. In an adult, the same type of loading would tend to cause a ligament tear since the epiphysis and diaphysis are fused (Pappas 1983) (Figure 4.22).

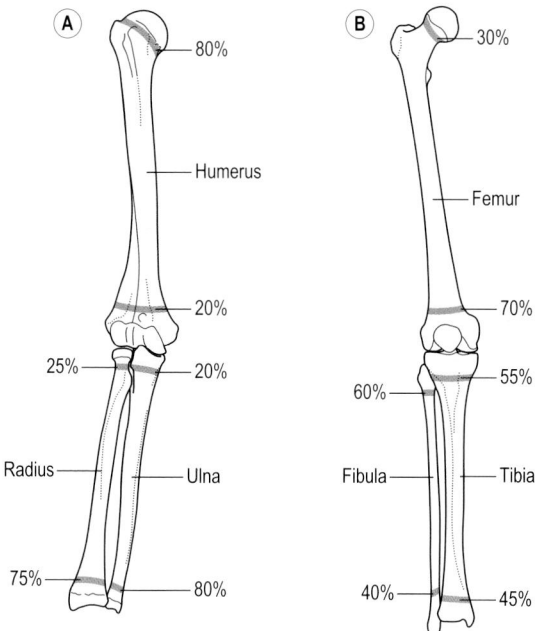

Figure 4.21 Contributions of the proximal and distal epiphyseal plates to growth in length of the long bones of the upper limb (A) and lower limb (B).

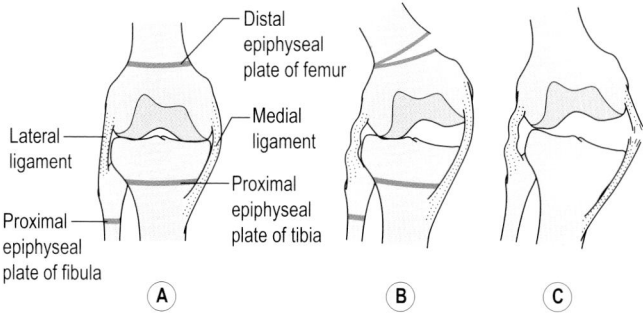

Figure 4.22 Effect of degree of skeletal maturity on type of injury. (A) Anterior aspect of the right knee joint showing normal alignment of the femur, tibia and fibula. (B, C) Excessive abduction of the knee joint. In a child this is more likely to result in a fracture through the distal epiphyseal plate of the femur than tearing of the medial ligament (B). After maturity, it is likely to result in partial or complete tearing of the medial ligament (C).

Growth of epiphyses

Just as the epiphyseal plates are responsible for growth in length of a bone, the hyaline cartilage that covers the end of a bone is responsible for growth of the epiphysis. This cartilage consists of an articular region and a non-articular region. Like an epiphyseal plate, articular cartilage (and its adjacent non-articular regions; see Figure 4.19) consists of four layers. The only real difference in structure between articular cartilage and an epiphyseal plate is in the arrangement of the fibres in the reserve layer. In an epiphyseal plate, the collagen fibres cross each other obliquely, forming a strong bond between the epiphyseal bone and the proliferation layer of the plate. In articular cartilage, the reserve layer is the outer layer. Whereas the majority of the layer is similar in structure to the reserve layer of an epiphyseal plate, the outer surface of articular cartilage is cell free and consists of densely packed collagen fibres and fibrils arranged parallel to the articular surface. This arrangement produces a tough wear-resistant surface.

The type of growth produced by articular cartilage is the same as that produced by an epiphyseal plate, i.e. interstitial growth. During the growth period, the rate of ossification of an epiphysis is greater than the rate of growth of the epiphysis. Consequently, the thickness of the articular cartilage becomes relatively thinner with age (Figure 4.23). At maturity, the thickness of articular cartilage is approximately 1 to 7 mm and this tends to decrease with age due to mechanical wear. Whereas bone growth is largely determined by genetic factors, the mechanical stress experienced by articular cartilage and epiphyseal plates, as a result of movement and the maintenance of an upright posture, also has a major effect on bone growth.

Growth of apophyses

Secondary centres of ossification occur not only in the epiphyses of long bones, but also in some of the rudimentary tuberosities of some bones, including the femur (greater trochanter and lesser trochanter), the innominate bones (ischial tuberosities, anterior superior and anterior inferior iliac spines) and the calcaneus (posterior calcaneal tuberosity) (Figures 4.24 and 4.25). The rudimentary tuberosities where these secondary centres of ossification occur (around 10 to 14 months after birth) are called apophyses. Each apophysis grows and ossifies in much the same way as an epiphysis. Apophyses provide areas of attachment for the tendons of powerful muscles such as the quadriceps (tibial tuberosity), hamstrings (ischial tuberosity) and calf muscles (calcaneal tuberosity) (see Figure 4.25). This form of attachment is different from that of most tendons, which attach directly onto the periosteum.

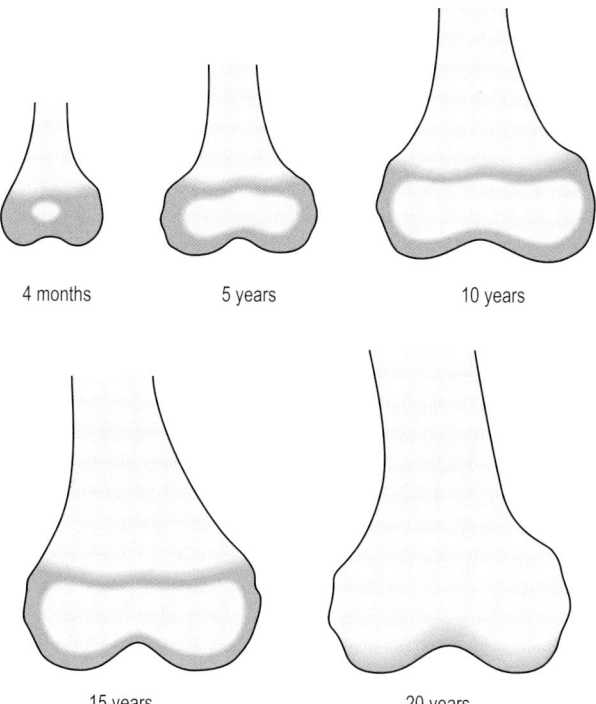

4 months   5 years   10 years

15 years   20 years

Figure 4.23 Successive stages in the ossification of the distal femoral epiphysis.

Prior to maturity each apophysis is separated from the rest of the bone by an apophyseal plate, which is very similar in structure and function to an epiphyseal plate. Each apophyseal plate is responsible for growth of the bone adjacent to the non-apophyseal side of the plate. Growth of the apophysis itself is due to a layer of cartilage (mixture of hyaline cartilage and fibrocartilage) outside of the apophysis into which the fibres of the tendon insert (see **Figure 4.24C**). At maturity, the apophyses fuse with the rest of the bone.

Epiphyses, especially those that form weight-bearing joints, are normally subjected to compression load and, as such, are often referred to as pressure epiphyses. In contrast, apophyses are normally subjected to tension load and are often referred to as traction epiphyses. Whereas apophyseal growth plates do not affect growth in bone length, they

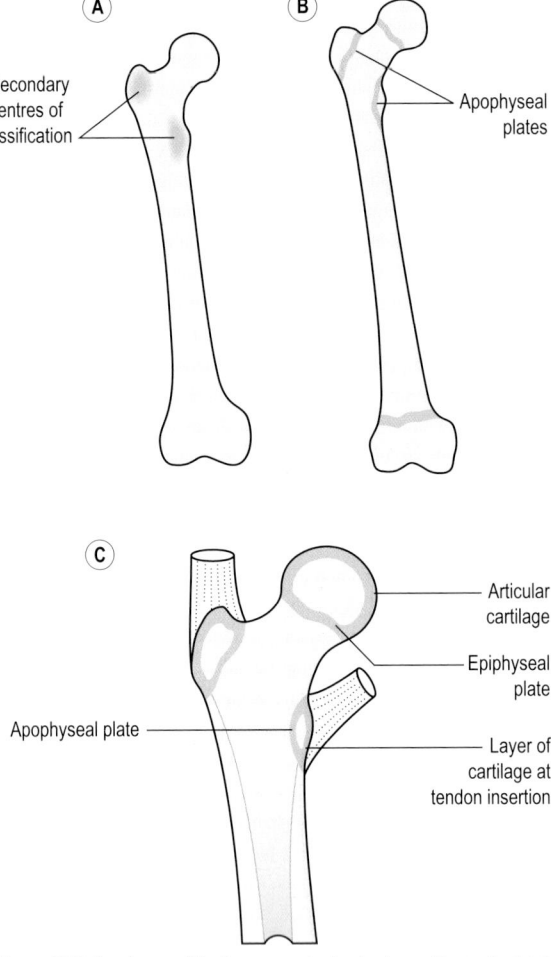

Figure 4.24 Apophyses of the femur: greater trochanter and lesser trochanter.
(A) Occurrence of secondary centres of ossification. (B) Apophyseal and epiphyseal plates of the femur. (C) Growth areas of the head of the femur and the greater and lesser trochanters.

do affect the alignment and strength of the tendons attached to them. Consequently, injury to apophyseal plates may affect the mechanical characteristics of associated muscles, which, in turn, may affect normal joint function.

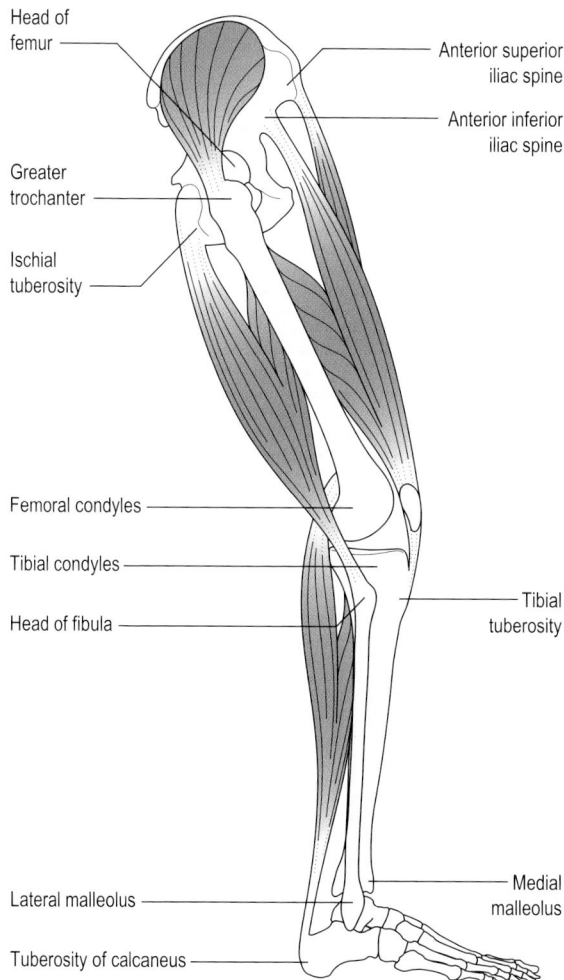

Head of femur

Anterior superior iliac spine

Anterior inferior iliac spine

Greater trochanter

Ischial tuberosity

Femoral condyles

Tibial condyles

Head of fibula

Tibial tuberosity

Lateral malleolus

Medial malleolus

Tuberosity of calcaneus

Figure 4.25 Major epiphyses (head of femur, femoral condyles, tibial condyles) and apophyses (anterior superior and anterior inferior iliac spines, greater trochanter, ischial tuberosity, lateral malleolus, medial malleolus, tuberosity of the calcaneus) of the lower limb.

Studies of sport-related injuries in children show that the proportion of injuries involving growth plates (epiphyseal and apophyseal) is between 6% and 18% of the total number of injuries (Speer & Braun 1985; Krueger-Franke et al 1992; Gross et al 1994). About 5% of these growth plate injuries result in some type of bone deformity (Larson 1973). On the basis of these figures, the number of growth-plate injuries resulting in bone deformity is in the region of 3 to 9 per thousand. However, this estimate is likely to be conservative since many injuries that occur during free play and sports are not reported or are incorrectly diagnosed (Combs 1994).

Structure of mature bone

Different regions of the bone are subjected to different types and magnitudes of loading. For example, the epiphyses are mainly subjected to compression loads, whereas the shaft is mainly subjected to bending and torsion loads. Not surprisingly, the structure of a mature bone reflects the normal loading pattern on the bone.

Compact and cancellous bone

Mature bone consists of osteones, also referred to as haversian systems (Figure 4.26). The only difference in structure in the different regions of a bone is in the extent to which the osteones are packed together; the closer the packing, the greater the density of the bone, the greater the strength of the bone and the lower the flexibility of the bone. In the hollow shaft of a long bone, the osteones are very closely packed together. This high-density bone is called compact bone (Figure 4.27).

A single osteone is a column of bone that consists of three to nine concentric rings (or layers) of bone surrounding a central open channel (see Figure 4.26). The concentric rings of bone are called lamellae and the

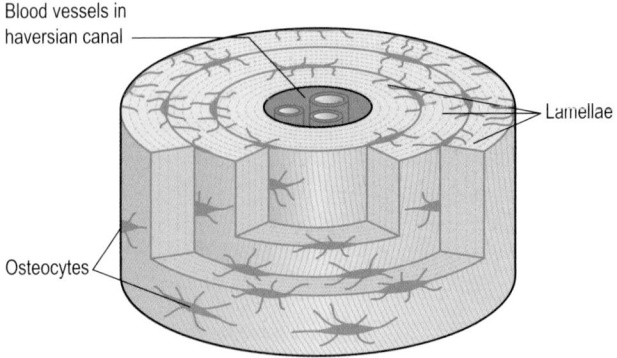

Figure 4.26 Structure of an osteone.

central channel is called a haversian canal. Each lamella consists of a single layer of closely packed collagen fibres arranged parallel to each other, embedded in bone salt. Whereas the collagen fibres in each lamella are parallel to each other, the orientation of fibres in adjacent lamellae is different. This arrangement, similar to the layers of collagen fibres in a joint capsule, enables the bone to strongly resist deformation in any direction.

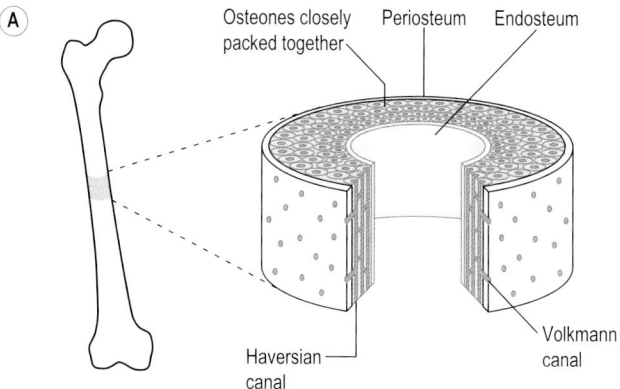

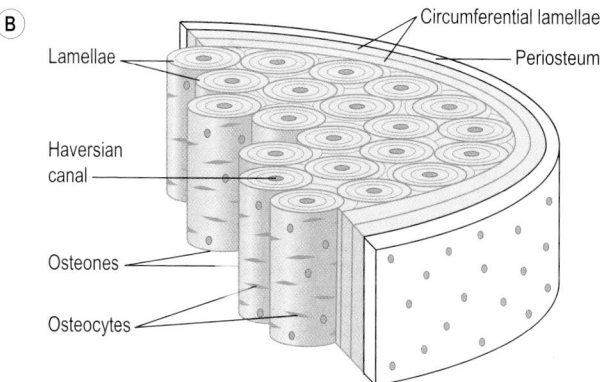

Figure 4.27 Structure of compact bone. (A) Section of the middle of the shaft of a mature long bone showing the network of haversian and volkmann canals. (B) Section of compact bone showing circumferential lamellae.

Between the lamellae are numerous osteocytes (Figure 4.28A). Osteocytes are basically osteoblasts that have become entrapped in the bone. They are mainly responsible for repairing bone damage, and are also involved in regulating the level of minerals, espccially calcium and phosphorus, in the blood (Bailey et al 1986). Each osteocyte lies in a lacuna, a small space, and the lacunae are linked together by tiny channels called canaliculi (see Figure 4.28B). The canaliculi run between the lamellae (circumferential canaliculi) and across the lamellae from one side to the other (radial canaliculi). This system of canaliculi enables the osteocytes to communicate with each other both physically, by means of projections from the cell bodies into the canaliculi, and chemically, by means of secretions from the cells. Communication between the osteocytes is important for coordinating the growth, development and repair of lamellae, and for facilitating intercellular exchange between osteocytes and between the osteocytes and blood vessels.

The haversian canal at the centre of each osteone contains blood vessels and nerves supported by loose connective tissue (see Figure 4.26). The canaliculi are linked to the haversian canals, thereby facilitating intercellular exchange between blood vessels and osteocytes. In addition to being linked together by canaliculi, the haversian canals of adjacent osteones are also linked together by channels called volkmann canals, which are similar in size to haversian canals (see Figure 4.27). Volkmann

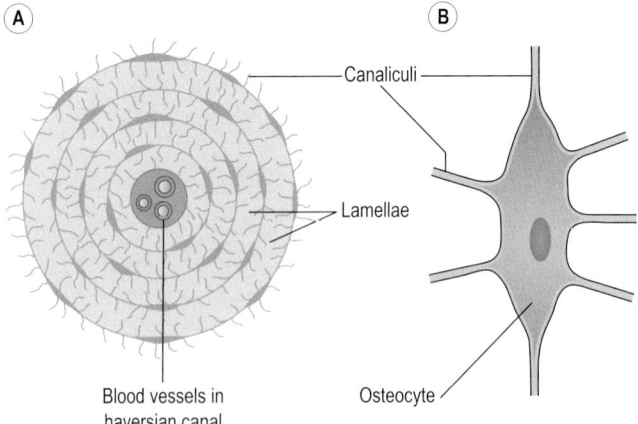

Figure 4.28 Osteocytes and canaliculi. (A) Cross section through an osteone showing circumferential and radial canaliculi. (B) An osteone lying in a lacuna with projections of the cell body into the canaliculi.

canals, like haversian canals, contain blood vessels and nerves supported by loose connective tissue. The volkmann canals form a system of channels, which run from the outside of the bone to the medullary cavity, by linking together the periosteum, haversian canals and endosteum (see Figure 4.27A). The system of haversian and volkmann canals enables blood vessels and nerves to pass along, around and across the bone.

The thickness of compact bone in the shaft decreases towards the epiphyses and the osteones start to separate into groups, rather like the branches of a tree separate from the trunk. However, in comparison with the branches of a tree, the groups of osteones, called trabeculae, form a distinct latticework pattern within each epiphysis (Figure 4.29). The latticework arrangement of the trabeculae is called trabecular bone, spongy bone or, most frequently, cancellous bone due to the large number of spaces between the trabeculae (cancellous = porous structure). The

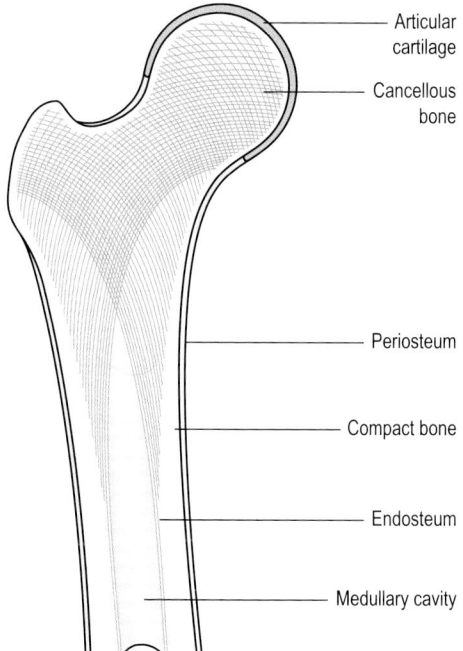

Articular
cartilage

Cancellous
bone

Periosteum

Compact bone

Endosteum

Medullary cavity

Figure 4.29 The structure of a mature long bone: a longitudinal section through the proximal third of the femur.

terminal branches of the trabeculae merge with a layer of calcified cartilage, called subchondral bone, adjacent to the articular cartilage.

The majority of the trabeculae cross each other at right angles; this arrangement maximizes the strength of the trabecular bone and minimizes the stress experienced by the epiphyses in all positions of the joint during habitual movements. The spaces between the trabeculae are filled with red marrow, i.e. loose connective tissue containing a large number of blood vessels, some white blood cells and fat cells, and a large number of cells called erythroblasts responsible for producing red blood cells. The spaces in cancellous bone are continuous with the medullary cavity and, therefore, the red marrow is continuous with the yellow marrow.

Cancellous bone is far less dense and, consequently, much more elastic than compact bone. The elasticity of cancellous bone is very important in ensuring congruity in joints during load transmission, thereby minimizing stress within the epiphyses and on the articular cartilages (Ascenzi & Bell 1971; Radin 1984).

Modelling and remodelling in bone

Growth, development and maintenance of bone are determined by the interaction of three subprocesses: skeletal genotype, modelling and remodelling. The skeletal genotype refers to the process of genetically programmed change in the external form (size and shape) and internal architecture of the bones. Modelling refers to the changes in the expression of the skeletal genotype that occur as a result of environmental factors such as nutrition and, in particular, the mechanical strains imposed by normal habitual activity. Remodelling refers to the coordination of osteoblastic and osteoclastic activity responsible for the actual changes in external form and internal architecture of the bones, including repair of bones (Figure 4.30). As bone is continuously being absorbed from some places (by osteoclasts) and deposited in others (by osteoblasts), the process of remodelling is sometimes referred to as turnover. Prior to maturity, all bones are in a continual state of change in external form and

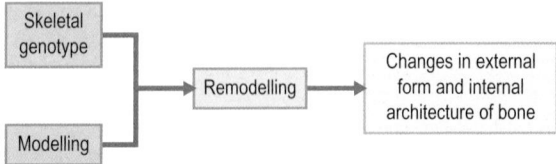

Figure 4.30 The relationship between skeletal genotype, modelling and remodelling in the growth, development and maintenance of bone.

internal architecture. After skeletal maturity is achieved (approximately 20 to 25 years of age), modelling of external form decreases to negligible proportions, but modelling of internal architecture continues throughout life (Frost 1979; Bailey 1995).

Porosity, osteopenia and osteoporosis

Owing to the various channels and spaces within compact and cancellous bone, any particular region of a bone consists of certain amounts of bone tissue and non-bone tissue. The term porosity describes the proportion of non-bone tissue. At skeletal maturity the porosity of compact and cancellous bone is approximately 2% and 50%, respectively; the density (amount of bone tissue per unit volume) of compact bone is approximately double that of cancellous bone (Radin 1984). The density of bone tissue depends on the degree of mineralization. During ossification, the degree of mineralization of bone tissue gradually increases and reaches a maximum level at skeletal maturity (Bailey et al 1986). However, the amount of bone within the skeleton may continue to increase for 5 to 10 years after skeletal maturity, especially in physically active individuals (Talmage & Anderson 1984; Stillman et al 1986). Consequently, bone mass peaks in males and females between 25 and 30 years of age. In terms of turnover, this means that from skeletal maturity to the age at which peak bone mass occurs, more new bone is formed than old and damaged bone is absorbed.

Following peak bone mass there is usually a stable period in which the amount of bone in the skeleton remains about the same, i.e. there is a balance between bone absorption and bone formation. This stable period is followed by a gradual decrease in bone mass for the rest of the life of the individual, i.e. the rate of bone absorption exceeds the rate of bone formation. Bone mass is the product of bone volume and bone density. The loss in bone mass that occurs with age following peak bone mass is the result of decreases in bone volume and bone density. Osteopenia refers to a level of bone density below the normal level for the age and sex of the individual (Bailey 1995).

Bone mass starts to decrease earlier and at a greater rate in females than in males. In males, bone loss normally starts to occur between 45 and 50 years of age and proceeds at a rate of 0.4% to 0.75% per year (Smith 1982; Bailey et al. 1986). In females, bone loss has three phases. The first phase starts around 30 to 35 years of age and proceeds at a rate of 0.75% to 1% per year until the menopause. From menopause until about 5 years after menopause the rate of bone loss increases to between 2% and 3% per year. During the final phase, the rate of bone loss is approximately 1% per year. On this basis, females may lose 40%

of their peak bone mass by the age of 80 years. In contrast, males may lose 20% of their peak bone mass by the age of 80 years (Figure 4.31).

Even though body weight tends to decrease with age, the rate of bone loss is usually much greater than the rate at which body weight decreases. Consequently, the effect of bone loss is that the bones, especially weight-bearing bones, become progressively weaker relative to the weight of the rest of the body. In addition to a gradual decrease in strength, the bones also gradually lose their elasticity (due to loss of collagen) and, as a result, become progressively more brittle. In some individuals, especially females, the loss of bone mass and elasticity may decrease to a level where some bones are no longer able to withstand the loads imposed by normal habitual activity. Consequently, these bones become very susceptible to fracture. This condition, the most common bone disorder in elderly people, is called osteoporosis (Bailey et al 1986). Osteoporosis may cause severe disfigurement, especially of the trunk, due to fractured or crushed vertebrae. Many deaths in the elderly are due to complications arising from bone fractures that occur as a result of osteoporosis (Kaplan 1983).

Bone loss tends to occur earlier and to proceed at a faster rate in cancellous bone than in compact bone (Bailey et al 1986). Consequently, regions of bones with a high proportion of cancellous bone, such as the bodies of the vertebrae, the head and neck of the femur, and the

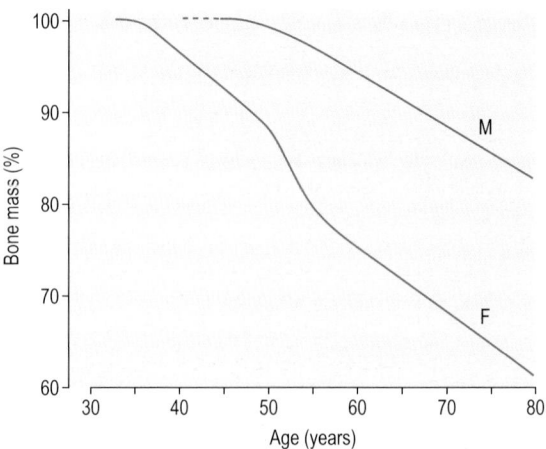

Figure 4.31 Effect of aging on bone mass. M = male; F = female.

distal end of the radius, are particularly vulnerable to osteoporosis and, therefore, to fracture in elderly people. This vulnerability is reflected in studies indicating a rapid increase in the incidence of bone fractures with age, especially in females. For example, the results of one study showed that the incidence of fracture to the distal end of the radius was seven times higher in 54-year-old women than in 40-year-old women (Bauer 1960). In another study, the incidence of fracture of the neck of the femur was found to be 50 times higher in 70-year-old women than in 40-year-old women (Chalmers & Ho 1970). With regard to compact bone, bone loss occurs mainly on the endosteal surface so that bone width remains relatively unchanged into old age (Smith 1982).

Whereas the cause of osteoporosis is not yet clear, there is general agreement that four variables have a major influence on the onset and progression of osteoporosis: genetic factors, endocrine status, nutritional factors and physical activity (Bailey et al 1986; MacKinnon 1988). The relative contribution of these variables has not yet been established, but physical activity level seems to have a major influence. In the absence of weight-bearing activity, no amount of endocrine or nutritional intervention will prevent rapid bone loss; there must be mechanical stress (Bailey et al 1986). Research suggests that regular moderate physical activity throughout life can help to prevent osteoporosis in three ways (Talmage & Anderson 1984; Bailey et al 1986; Stillman et al. 1986; Smith & Gilligan 1987):

1. Peak bone mass is directly related to the level of physical activity prior to peak bone mass; the higher the peak bone mass, the lower the risk of osteoporosis (Rautava et al 2007).
2. An above-average level of physical activity after peak bone mass will delay the onset of bone loss (Kemmler et al 2007; Maddalozzo et al 2007).
3. An above-average level of physical activity after peak bone mass will reduce the rate of bone loss (Kemmler et al 2007; Maddalozzo et al 2007).

Key Concepts

Many individuals, especially females, develop osteoporosis. The cause of osteoporosis is not yet clear, but lack of mechanical stress appears to be a major influence. Regular physical activity throughout life appears to be the best way of preventing osteoporosis

Structural adaptation of the musculoskeletal system

In any body position other than the relaxed recumbent position, the musculoskeletal system is likely to be subjected to considerable loading. In response to the forces exerted on them, the musculoskeletal components experience strain, i.e. they are deformed to a certain extent, and the greater the force, the greater the strain. Under normal circumstances the musculoskeletal components adapt their external form (size and shape) and internal architecture (structure) to the time-averaged forces exerted on them in order to more readily withstand the strain (Carter et al 1991). However, when the degree of strain experienced by a particular component exceeds its strength, it becomes injured. Consequently, there is an intimate relationship between the structure and function of the musculoskeletal system (Watkins 1999).

Structural adaptation in bone

The last 30 years have produced much of the present knowledge concerning the adaptation of musculoskeletal components to changes in time-averaged load (Frost 1988a, 1988b, 1990, 2004). However, the fundamental concepts concerning the adaptation of bone were established over 100 years ago (Gross & Bain 1993). In 1892 Julius Wolff (1836–1902) summarized the contemporary views of bone adaptation to changes in time-averaged load in what came to be known as Wolff's law (Wolff 1988). Wolff's law, shown to be more or less correct, hypothesized that bone adapts its external form and internal architecture to the time-averaged load exerted on it in an ordered and predictable manner to provide optimal strength with minimal bone mass.

The adaptation of bone to time-averaged load is referred to as modelling. In normal growth and development, modelling has been estimated to account for 20% to 50% of the dimensions of mature bones (Frost 1988b). Some of the load experienced by bone is due to the weight of body segments. However, this source of loading is small relative to the loads exerted by muscles (Watkins 1999). From birth to maturity, bone has the capacity to model external form and internal architecture. However, the capacity to model external form gradually decreases and virtually ceases at maturity. The capacity to model internal architecture also decreases with age, but is retained to some extent throughout life. In general, bone adapts to changes in time-averaged loads by increasing or decreasing bone mass to maintain an optimum strain environment. In bone, the optimum strain environment is characterized by minimal flexure (or bending) strain and an even distribution of stress (usually compression stress) across articular areas. An even distribution of stress across articular areas is maintained by modelling in accordance with the phenomenon of chondral modelling (Frost 1979).

The chondral modelling phenomenon

All bones that develop from hyaline cartilage via endochondral ossification experience chondral modelling, i.e. the rate and amount of new bone formed by hyaline cartilage depends upon the amount and form of load exerted on it. Chondral modelling applies to articular cartilage, epiphyseal plates, insertions of tendons and ligaments, apophyseal plates, end plates in symphysis joints and sesamoid bones (Watkins 1999).

In a long bone, the size and shape of the epiphyses and metaphyses, and, consequently, the orientation of the epiphyses of a bone to its shaft, are determined by chondral modelling in articular cartilage and epiphyseal plates. The area of contact between the articular surfaces in a synovial joint varies with the angle of the joint. However, irrespective of the angle of the joint, the articular surfaces are normally congruent, i.e. the compression stress on the articular cartilages is evenly distributed. Incongruence results in an unequal distribution of load across articular cartilages and epiphyseal plates. If prolonged, such unequal loading results in modelling to restore normal congruence. However, the actual changes that occur depend on the extent of the changes in the patterns of loading on the articular cartilage and epiphyseal plates. If the changes in loading remain within the normal range, then a negative-feedback mode of modelling is invoked resulting in restoration of normal congruence with normal or slightly abnormal alignment of the bones. However, if the changes in loading are outside the normal range, then a positive-feedback mode of modelling is invoked, which aggravates the condition, resulting in progressively worsening misalignment.

Modelling of metaphyses and epiphyses

A functionally normal joint is a congruent joint that transmits loads across the articulating surfaces in a normal manner. An anatomical misalignment at the knee, or any other joint, will be functionally normal if the misalignment stabilizes (does not get progressively worse). In these cases, the anatomical misalignments represent normal modelling in response to abnormal patterns of loading. The skeletal adaptations ensure normal transmission of loads across the joints. Figure 4.32 illustrates the effect of negative-feedback in relation to abductor/adductor muscle imbalance at the knee. Figure 4.32A represents a knee with normal balance between the

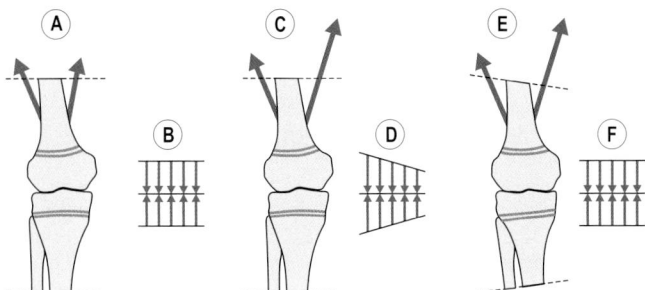

Figure 4.32 Modelling of the metaphyses and epiphyses: effect of the negative feedback mode in relation to an abductor/adductor imbalance at the knee. (A and B) Abductor/adductor balance at the knee (A) resulting in an even distribution of load across the articular surfaces and epiphyseal plates (B). (C and D) Abductor/adductor imbalance at the knee (C) resulting in an uneven distribution of load across the articular surfaces and epiphyseal plates (D). (E and F) Negative feedback adaptive growth of epiphyses and metaphyses in response to the abductor/adductor imbalance (E) that results in an even distribution of load across the articular surfaces and epiphyseal plates (F).

abductor and adductor muscles, i.e. the resultant horizontal force at the knee is zero. This situation is associated with normal alignment between the femur and tibia and an even distribution of load across the articular surfaces and epiphyseal plates (see **Figure 4.32B**). **Figure 4.32C** shows the same knee with an abductor-adductor imbalance such that there is a net medially directed horizontal force at the knee tending to increase the degree of genu valgum. **Figure 4.32D** shows the unequal pattern of loading on the articular surfaces and epiphyseal plates associated with the muscle imbalance. Assuming that the unequal loading is within the normal range, the negative-feedback mode is invoked. The rate of growth of the lateral aspects of the epiphyses and metaphyses is increased and the rate of growth of the medial aspects of the epiphyses and metaphyses is decreased such that normal congruence is restored (with net zero horizontal force at the knee) at the expense of an abnormal alignment between the femur and tibia, i.e. much reduced genu valgum or even slight genu varum relative to most individuals (see **Figure 4.32E and F**).

Whether or not a particular joint is anatomically misaligned during childhood, the only time it may become painful (excluding injuries and pathological conditions not due to loading) is during adulthood, when

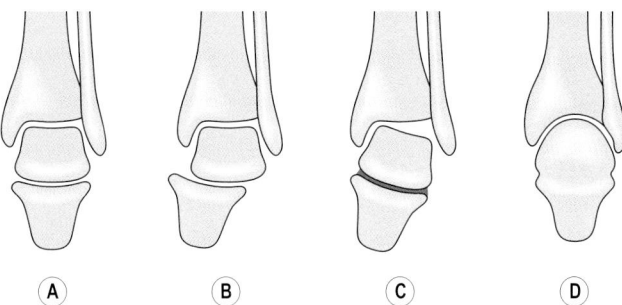

Figure 4.33 Modelling of articular surface. (A) Normal orientation of the tibia, fibula, talus and calcaneus in the neutral position. (B) Normal movement of the calcaneus with respect to the talus during inversion. (C) Abnormal movement of the talus within the tibiofibula mortise during eversion of the calcaneus as a result of restricted movement in the subtalar joint. (D) Adaptive growth of the articular surfaces of the ankle joint resulting in a ball and socket ankle joint rather than a hinge joint.

the bones are no longer capable of modelling in response to abnormal loading. In most adults, abnormal patterns of loading are the result of an increasingly sedentary lifestyle in which body weight gradually increases and muscle strength gradually decreases.

Modelling of articular surfaces

Minor incongruences between articular surfaces in synovial joints tend to result in large changes in the compression stress experienced by different parts of the articular surfaces. This is especially the case in joints with pulley-shaped articular surfaces such as the ankle joint (Figure 4.33). Under normal circumstances, inversion and eversion of the foot take place about the subtalar joint (see Figure 4.33A and B). However, if movement at the joint is absent or limited, inversion and eversion of the foot twists the talus in the tibiofibula mortise resulting in excessive compression stress on those parts of the articular surfaces that remain in contact (see Figure 4.33C). The excessive loading on the impinging areas reduces or halts growth in these areas, while growth of the unloaded areas proceeds at the normal rate. Consequently, the shapes of the articular surfaces adapt to the abnormal loading conditions by forming a rounded surface in the coronal plane rather than a pulley-shaped surface, and the ankle joint as a whole resembles a ball and socket joint rather than a hinge joint (see Figure 4.33D) (Frost 1979).

Review questions

1. Describe the two main functions of connective tissues.
2. Differentiate between the following:
   - Regular and irregular ordinary connective tissues
   - Loose connective tissue and adipose connective tissue
   - Intracapsular and extracapsular ligaments
   - Ligaments and retinacula
   - Fibrous tissue and elastic tissue
3. Describe the different types of cell found in ordinary connective tissues.
4. Describe the two main functions of cartilage.
5. Differentiate between elasticity and viscoelasticity.
6. Explain why repair in cartilage is usually slow and may not take place at all.
7. Describe the three main mechanical functions of bone.
8. Differentiate between the following:
   - Primary and secondary centres of ossification
   - Appositional and interstitial growth
   - Epiphyseal and apophyseal plates
   - Osteoblasts, osteoclasts and osteocytes
   - Compact and cancellous bone
   - Osteopenia and osteoporosis
   - Anatomical and functional alignment in normal joints

References

Akeson WH, Frank CB, Amiel D, Woo SLY (1985) Ligament biology and biomechanics. In: Finerman G (ed) Symposium on sports medicine: the knee. Mosby, St Louis, pp 111–15.

Alexander RNcN (1968) Animal mechanics. Sidgwick & Jackson, London.

Alexander RNcN (1975) Biomechanics. Chapman & Hall, London.

Ascenzi A, Bell GH (1971) Bone as a mechanical engineering problem. In: Boume GH (ed) The biochemistry and physiology of bone. Academic Press, New York, pp 41–57.

Bailey DA (1995) The role of mechanical loading in the regulation of skeletal development during growth. In: Blimkie CJR, Bar-Or O (eds) New horizons in pediatric exercise science. Human Kinetics, Champaign, IL, pp 97–108.

Bailey DA, Martin AD, Houston CS, Howie JL (1986) Physical activity, nutrition, bone density and osteoporosis. Australian Journal of Science and Medicine in Sport 18:3–8.

Bauer G (1960) Epidemiology of fracture in aged persons. Clinical Orthopedics 17:219–25.

Caplan AI (1984) Cartilage. Scientific American 251:82–90.

Carter DR, Wong M, Orr TE (1991) Musculoskeletal ontogeny, phylogeny, and functional adaptation. Journal of Biomechanics 24S1:3–16.

Chalmers J, Ho KC (1970) Geographical variations in senile osteoporosis. Journal of Bone and Joint Surgery 52:667–75.

Combs JA (1994) Hip and pelvis avulsion fractures in adolescents. The Physician and Sportsmedicine 22:41–44, 47-9

Frost HM (1979) A chondral modeling theory. Calcified Tissue International 28:181–200.

Frost HM (1988a) Structural adaptations to mechanical usage: a proposed three-way rule for bone modelling. Part I. Veterinary and Comparative Orthopaedics and Traumatology 1:7–17.

Frost HM (1988b) Structural adaptations to mechanical usage: a proposed three-way rule for bone modelling. Part II. Veterinary and Comparative Orthopaedics and Traumatology 2:80–85.

Frost HM (1990) Skeletal structural adaptations to mechanical usage: four mechanical influences on intact fibrous tissues. The Anatomical Record 226:433–9.

Frost HM (2004) A 2003 update of bone physiology and Wolff's law for clinicians. Angle Orthodontist 74:3–15.

Gross ML, Flynn M, Sonzogni JJ (1994) Overworked shoulders: managing injury of the proximal humeral epiphysis. The Physician and Sportsmedicine 22:81–82, 85-6

Gross TS, Bain ST (1993) Skeletal adaptation to functional stimuli. In: Grabiner MD (ed) Current issues in biomechanics. Human Kinetics, Champaign, IL, pp 151–69.

Kaplan FS (1983) Osteoporosis. Clinical Symposia. Ciba-Geigy 35:32–42.

Kemmler W, Engelke K, von Stengel S, et al (2007) Long-term four-year exercise has a positive effect on menopausal risk factors: the Erlangen Fitness Osteoporosis prevention Study. Journal of Strength and Conditioning Research 21:232–9.

Krueger-Franke M, Siebert CH, Pfoerringer W (1992) Sports-related epiphyseal injuries of the lower extremity: an epidemiological study. Journal of Sports Medicine and Physical Fitness 32:106–11.

Larson RL (1973) Physical activity and the growth and development of bone and joint structures. In: Rarick GL (ed) Physical activity: human growth and development. Academic Press, New York.

Larson RL, McMahan RO (1966) The epiphyses and the childhood athlete. Journal of the American Medical Association 196:607–12.

MacKinnon JL (1988) Osteoporosis. Physical Therapy 10:1533–9.

Maddalozzo GF, Widrick JJ, Cardinal BJ, et al (2007) The effects of hormone replacement therapy and resistance training on spine bone mineral density in early postmenopausal women. Bone 40:1244–51.

McArdle WD, Katch FI, Katch VL (1996) Exercise physiology: energy, nutrition, and human performance. Lea & Febiger, Philadelphia.

Nordin MN, Frankel VH (1989) Basic biomechanics of the skeletal system. Lea & Febiger, Philadelphia.

Pappas AM (1983) Epiphyseal injuries in sports. The Physician and Sportsmedicine 11:140–8.

Peterson HA (2001) Physeal injuries and growth arrest. In: Beaty JH, Kasser JR (eds) Fracture in children. Lippincott Williams & Wilkins, Philadelphia, pp 91–138.

Radin EL (1984) Biomechanical considerations. In: Moskowitz RW, Howell DS, Goldberg VM, Mankin HJ (eds) Osteoarthritis: diagnosis and management. Saunders, Philadelphia, pp 71–9.

Rautava E, Lehtonen-Veromaa M, Kautiainen H, et al (2007) The reduction of physical activity reflects on the bone mass among young females: a follow-up study of 142 adolescent girls. Osteoporosis International 18:915–22.

Smith EL (1982) Exercise for prevention of osteoporosis. The Physician and Sportsmedicine 10:72–83.

Smith EL, Gilligan C (1987) Effects of inactivity and exercise on bone. The Physician and Sportsmedicine 15:91–100.

Speer DP, Braun JK (1985) The biomechanical basis of growth plate injuries. The Physician and Sportsmedicine 13:72–8.

Stillman RJ, Lohman TG, Slaughter MH, Massey BH (1986) Physical activity and bone mineral content in women aged 35 to 85 years. Medicine and Science in Sports and Exercise 18:576–80.

Talmage RV, Anderson JJB (1984) Bone density loss in women: effects of childhood activity, exercise, calcium intake, and oestrogen therapy. Calcified Tissue International 36:522–32.

Tortora GJ, Anagnostakos NP (1984) Principles of anatomy and physiology. Harper & Row, New York.

Watkins J (1999) Structure and function of the musculoskeletal system. Human Kinetics, Champaign, IL.

Williams PL, Bannister LH, Berry MM, et al, eds (1995) Gray's anatomy. Longman, Edinburgh.

Wolff J (1988) The law of bone modelling, Translated by P. Maquet and R. Furlong Springer Verlag, New York. Originally published as Wolff J (1892) Das gesetz der transformation der knochen. A. Hirschwald, Berlin.

CHAPTER 5

The articular system

The open-chain arrangement of the bones and the range of movement in the joints between the bones facilitates a very wide range of body postures. However, this movement capability is only possible at the expense of low mechanical advantage of the skeletal muscles and, in turn, high joint reaction forces. Consequently, joints are designed to allow movement while transmitting relatively large forces. The purpose of this chapter is to describe the structure and function of the various types of joints.

Structural classification of joints

A joint, also referred to as an articulation or an arthrosis, is defined as a region where two or more bones are connected. The adult skeleton normally has 206 bones linked by approximately 320 joints. The articular system refers to all of the joints of the body. Joints have two main functions: to facilitate relative motion between bones and to transmit forces from one bone to another.

In terms of structure, there are basically two types of joint:

1. Joints in which the articular (opposed) surfaces of the bones are united by either fibrous tissue or cartilage are called fibrous joints and cartilaginous joints, respectively (Figure 5.1A).
2. Joints in which the articular surfaces are not attached to each other but are held in

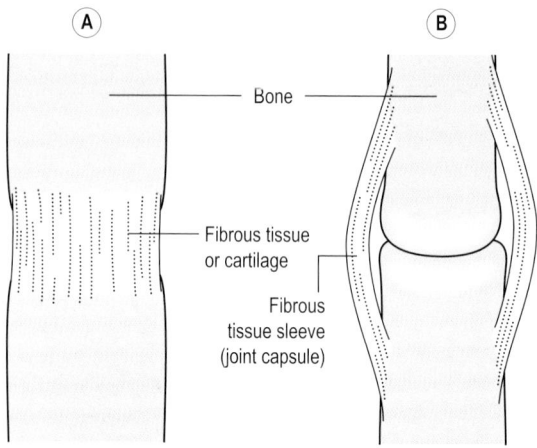

Figure 5.1 Two basic types of structure of joints. (A) Articular surfaces united by fibrous tissue or cartilage. (B) Articular surfaces not attached to each other but held in contact with each other by a fibrous tissue sleeve.

contact with each other by a sleeve of fibrous tissue supported by ligaments (**Figure 5.1B**) are referred to as synovial joints, and the fibrous sleeve is the joint capsule (see Chapter 4).

Fibrous joints

Fibrous joints are often referred to as syndesmoses (syn = with, desmo = ligament). The degree of movement in a syndesmosis is largely determined by the amount of fibrous tissue between the articular surfaces. In general, the smaller the amount of fibrous tissue, the lower the range of movement. There are two types of syndesmoses: membranous and sutural.

Membranous syndesmoses

In a membranous syndesmosis the articular surfaces are united by a sheet of fibrous tissue called an interosseous membrane (inter = between, osseous = bone). The interosseous membrane functions rather like webbing; it forms a flexible but fairly inextensible link between the articular surfaces. The radius and ulna are connected by an interosseous membrane (**Figure 5.2A**). The majority of the fibres in the membrane run obliquely downward and medially from the medial border of the radius to

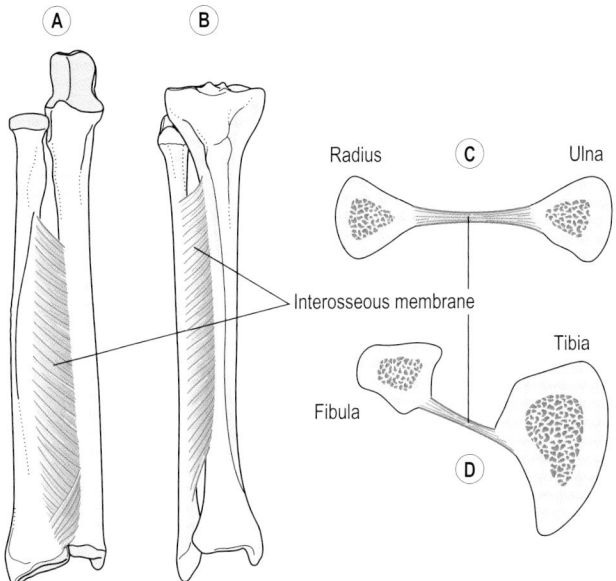

Figure 5.2 Anterior aspects and cross-sectional views of the membranous syndesmoses between the right radius and ulna (A and C) and the right tibia and fibula (B and D).

the lateral border of the ulna. The remaining fibres run obliquely downward and laterally from the lateral border of the ulna to the medial border of the radius. The interosseous membrane between the radius and ulna has two main functions: to stabilize the bones in all positions of the forearm, and to provide areas of attachment for muscles on the anterior and posterior aspects of the lower arm (see **Figure 5.2C**). As in the lower arm, there is an interosseous membrane in the lower leg that connects the medial border of the shaft of the fibula and the lateral border of the shaft of the tibia (see **Figure 5.2B**). The interosseous membrane between the fibula and tibia stabilizes the bones in all positions of the ankle joint, and provides areas of attachment for muscles on the anterior and posterior aspects of the lower leg (see **Figure 5.2D**).

Whereas the interosseous membranes in the lower arms and lower legs are permanent features, there is a particular group of syndesmoses, the sutures and fontanels of the skull (see **Figure 3.6A and C**), which originate as membranous syndesmoses and then change to sutural syndesmoses.

Sutural syndesmoses

In a sutural syndesmosis, the articular surfaces are united by a very thin layer of fibrous tissue. By late childhood all of the sutures and fontanels of the skull are converted from membranous to sutural syndesmoses (see **Figure 3.6**). The thin layer of fibrous tissue, together with close interlocking of the articular surfaces, tends to severely restrict movement in these joints. With increasing age, the sutures usually undergo ossification and, as such, are converted to synostoses (syn = with, osteo = bone).

Cartilaginous joints

The degree of movement in cartilaginous joints is determined by the type and thickness of the cartilage. There are two kinds of cartilaginous joints: synchondroses and symphyses.

Synchondroses

In a synchondrosis (syn = with, chondro = cartilage), the articular surfaces are united by hyaline cartilage. There are two types of synchondroses: temporary and permanent. The temporary synchondroses, sometimes referred to as physeal joints, include the following:

1. Epiphyseal plates, i.e. the regions of hyaline cartilage between the epiphyses and shaft of a long bone prior to maturity (see **Figure 4.21**).
2. Apophyseal plates, i.e. the region of hyaline cartilage between an apophysis and the rest of a bone prior to maturity (see **Figure 4.24**).
3. The regions of hyaline cartilage in an immature innominate bone between the ilium, ischium and pubis regions of the bone.

All temporary synchondroses are converted to synostoses at maturity. A few synchondroses, the permanent synchondroses, remain moderately flexible throughout life. These joints include the joints between the anterior ends of the ribs and the sternum (see **Figure 3.13**).

Symphyses

In a symphysis (sym = with, physeal = growth plate) the articular surfaces are united by a combination of hyaline cartilage and fibrocartilage. A layer of hyaline cartilage covers each articular surface, and sandwiched between the layers of hyaline cartilage is a relatively thick piece of fibrocartilage (**Figure 5.3**). The fibrocartilage is often referred to as a disc even though it is usually kidney shaped or oval. The joint is normally supported by a number of ligaments that cross the outside of the joint and attach onto the periphery of the fibrocartilage. As in all types of fibrous and cartilaginous joints, the bone, hyaline cartilage, and fibrocartilage regions in a symphysis joint are intimately connected; there is a gradual change

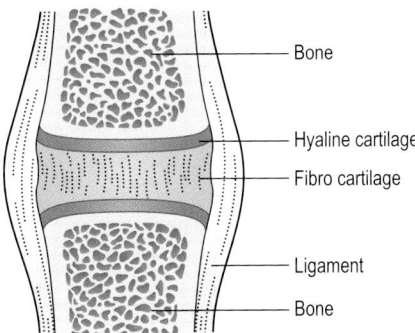

Figure 5.3 A typical symphysis joint.

from one region to another. In effect, the joint consists of a single piece of material whose flexibility varies across the joint.

Fibrocartilage readily deforms in response to bending and torsion loads. The degree of movement in a symphysis joint is largely determined by the thickness of the fibrocartilage; the thicker the fibrocartilage, the greater the flexibility. The joints between the bodies of the vertebrae (intervertebral joints) and the pubic symphysis are symphysis joints. Symphysis joints tend to remain moderately flexible throughout life.

Synovial joints

In the adult skeleton, approximately 80% of the joints are synovial (syn = with, vial = cavity). In general, synovial joints have greater range of movement than fibrous and cartilaginous joints. The body's capacity to adopt a broad range of postures is due largely to the range of movement in synovial joints. Virtually all of the joints in the upper and lower limbs are synovial.

In a synovial joint, each articular surface is covered with a layer of articular (hyaline) cartilage. The surfaces are not attached to each other but, under normal circumstances, are held in contact with each other, in all positions of the joint, by a joint capsule and various ligaments (**Figure 5.4**). During movement of a synovial joint the articular surfaces slide and roll on each other. The capsule encloses a joint cavity that, due to the close contact between the articular surfaces, is normally very small. In **Figure 5.4**, the joint cavity is shown much larger than normal to differentiate the features of the joint. The inner wall of the capsule and the non-articular bony surfaces inside the joint are covered with synovial membrane. Synovial membrane consists of areolar tissue (see Chapter 4) with specialized cells

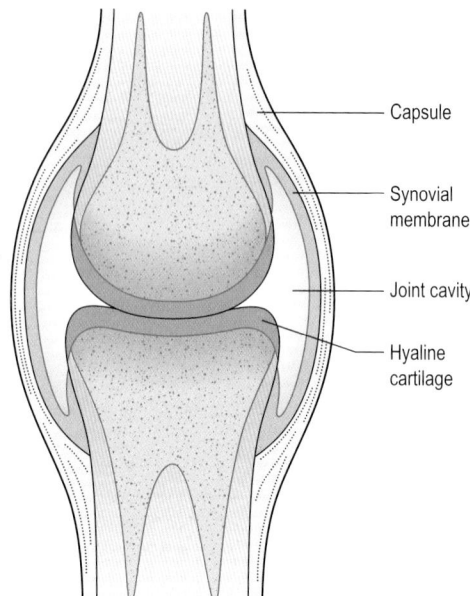

Figure 5.4 A typical synovial joint.

Labels: Capsule, Synovial membrane, Joint cavity, Hyaline cartilage

that secrete synovial fluid into the joint cavity. Synovial fluid is viscous and resembles raw egg white. It has two important functions:

1. A mechanical function: the fluid lubricates the articular surfaces so that they slide over each other easily, thereby preventing excessive wear.
2. A physiological function: the fluid seeps into the articular cartilage and nourishes the cartilage cells.

Key *Concepts*

In terms of structure, there are basically two types of joint:

- Joints in which the articular surfaces are united by either fibrous tissue or cartilage (fibrous joints and cartilaginous joints)
- Joints in which the articular surfaces are not attached to each other, but are held in contact with each other by a joint capsule and ligaments (synovial joints)

Congruence, articular discs and menisci

The articular surfaces in most synovial joints are reciprocally shaped, which normally results in a large contact area (relative to the area of the articular surfaces) between the opposed articular surfaces in all positions of the joint. For any particular joint position, the larger the contact area between the articular surfaces (the larger the area over which the joint reaction force is transmitted), the lower the compressive stress on the articular surfaces, and vice versa. Some synovial joints, such as the tibiofemoral joint, do not have reciprocally shaped articular surfaces so that, in the absence of other structures, the area of contact between the articular surfaces in any particular joint position would be relatively small and the compressive stress very high (see **Figure 3.23**). However, in such joints the effective area of contact between the articular surfaces is normally as large as in joints with reciprocally shaped articular surfaces due to the presence of fibrocartilaginous wedges between the unopposed parts of the articular surfaces; the fibrocartilaginous wedges distribute the joint reaction force over a large area of the articular surfaces. The parts of the fibrocartilagenous wedges that distribute the joint reaction force deform in response to the pressure of the joint reaction force. The wedges are normally held in position by attachment to the inner wall of the joint capsule or by attachment to bone adjacent to the articular surface (see **Figure 5.5**).

In the acromioclavicular joint, and sometimes the ulna-carpal joint, there is a single fibrocartilage wedge in the form of a ring that tapers from the outside towards the centre (**Figure 5.5A**). In the tibiofemoral joint there are normally two C-shaped fibrocartilage wedges; each wedge is called a meniscus (half-moon, crescent shape; **Figure 5.5B**). In some joints, such as the sternoclavicular and ulna-carpal joints, there is usually a complete disc of fibrocartilage that effectively divides the joint into two joints (see **Figure 5.5C and D**). A complete disc of fibrocartilage is referred to as an articular disc. The congruence of a joint refers to the area over which the joint reaction force is transmitted; in any particular joint position, the greater the area, the greater the congruence of the joint, and vice versa. Normally, congruence is maximized in order to minimize compressive stress on the articular surfaces. One of the main functions of articular discs and menisci is to improve congruence in joints. Improved congruence will tend to:

1. Reduce compressive stress on the opposed articular surfaces.
2. Help to maintain normal joint movements and effective distribution of synovial fluid over the articular surfaces.
3. Improve shock absorption.

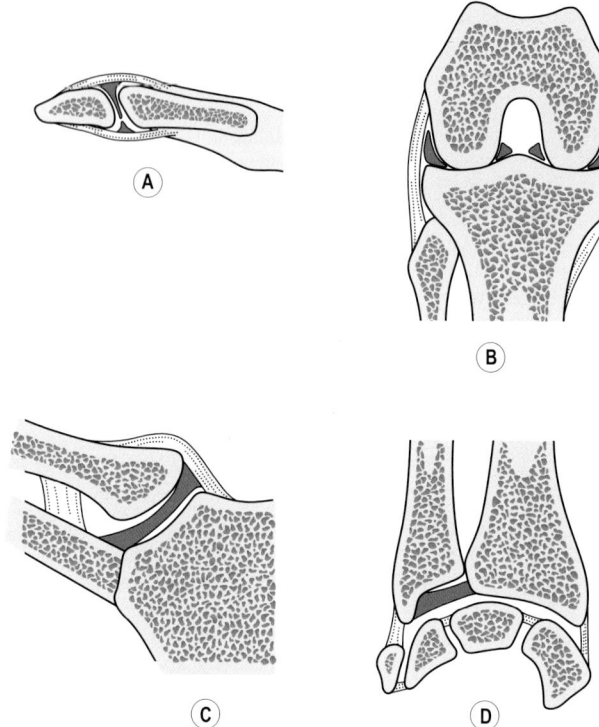

Figure 5.5 Articular discs and menisci. (A) Coronal section through the right acromioclavicular joint. (B) Coronal section through the right tibiofemoral joint with the joint flexed at approximately 90°. (C) Coronal section through the right sternoclavicular joint. (D) Coronal section through the left wrist joint.

Damage to articular discs and menisci (usually as a result of tearing) will adversely affect congruence and, in turn, increase the rate of joint degeneration. Whereas most joints fit exclusively into one of the main categories of joint (fibrous, cartilaginous, synovial), some joints have or develop characteristics of more than one category. Consequently, in terms of structure and function, the range of joints is very broad. This range reflects the capacity of the skeletal system for structural adaptation, i.e. the capacity to modify the structure of joints in relation to their particular functional requirements.

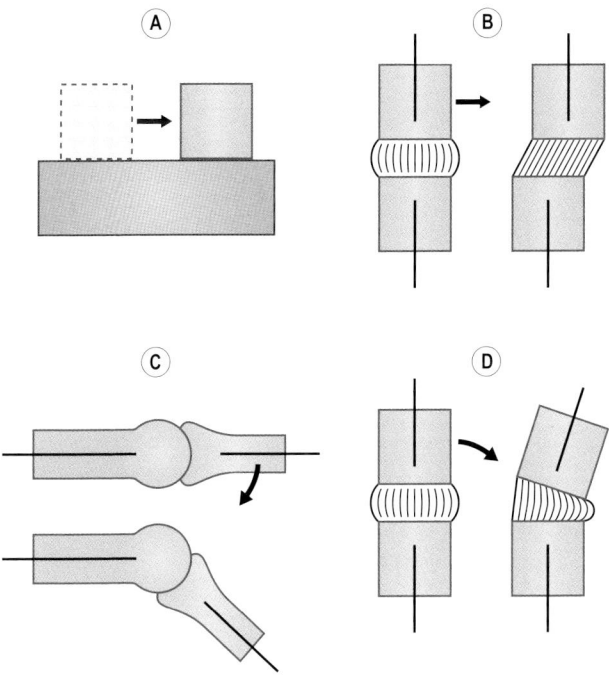

Figure 5.6 Types of joint movements. (A) Linear motion in a synovial joint. (B) Linear motion in a fibrous joint. (C) Angular motion in a synovial joint. (D) Angular motion in a fibrous joint.

Joint movements

When joint movement occurs, the dominant form of movement is normally angular. Linear movement may occur, but in normal joint movements the degree of linear movement is usually quite small. For example, the pubic symphysis and the intervertebral symphyses may experience a certain degree of linear movement similar to that shown in **Figure 5.6B**. When this happens the fibrocartilage in the joint is subjected to a shear load that is likely to tear the cartilage if the shear strain is beyond a very small amount. A limited amount of linear movement may occur in some synovial joints such as the intercarpal and intertarsal joints, and the joints between the superior and inferior articular facets of the vertebrae. However, in most synovial joints, linear movement is normally slight (Basmajian 1970). Linear movement beyond a very small amount almost certainly results in partial separation of the articular surfaces and damage to the capsule and ligaments of the joint.

Degrees of freedom

Figure 5.7 shows the position of the reference axes (see **Figure** 3.3) in relation to the shoulder joint. With respect to the reference axes, there are six possible directions, called degrees of freedom, in which the shoulder joint, or any other joint, might be able to move, depending upon its structure. The six directions consist of three linear directions (along the axes) and three angular directions (around the axes). A joint with six degrees of freedom could move in any direction by a combination of linear and angular movements. Some cartilaginous joints have six degrees of freedom, albeit within a small range of movement. In contrast, the larger synovial joints tend to have no linear degrees of freedom, but they usually have one to three angular degrees of freedom with a relatively large range of movement.

Most movements in everyday life such as walking, bending and reaching involve simultaneous or sequential movement in two or more joints. In such multi-joint movements, the number of degrees of freedom in that

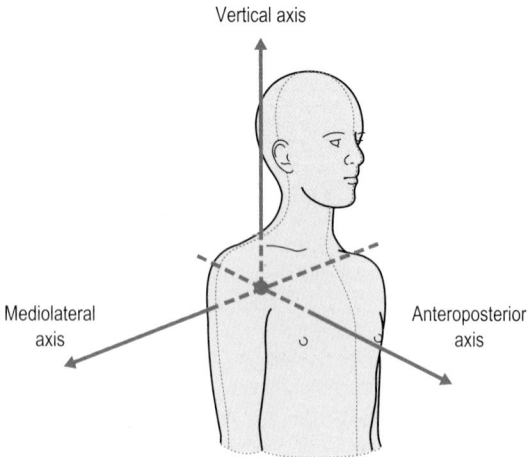

Figure 5.7 Reference axes with respect to the shoulder joint.

part of the skeletal system responsible for the movement is the sum of the degrees of freedom of the individual joints involved. Consequently, there is an almost infinite number of combinations of joint movements that could be employed in all multi-joint movements. Furthermore, temporary or permanent impairment in one joint can usually be compensated by a change in the movement of other joints.

Angular movements

Angular movements in joints refer to rotations around the three reference axes. With the anatomical position as a reference position, special terms describe the various angular movements.

In most joints, the terms abduction and adduction refer to rotations around the anteroposterior axis. In the shoulder, wrist and hip joints, abduction and adduction refer to movement of the arm, hand and leg away from and towards the median plane, respectively (**Figure 5.8A–C**). In the hand and foot, abduction occurs when the fingers and toes are spread, and adduction occurs when the fingers and toes are returned to the reference position (see **Figure 5.8D**).

In most joints, the terms flexion and extension refer to rotation around the mediolateral axis. In the shoulder, wrist and hip joints, flexion refers to movement of the arm, hand and leg forward, and extension refers to movement of the arm, hand and leg backward (**Figure 5.9A–C**). In the elbow, knee, metacarpophalangeal and interphalangeal (hand and foot) joints, flexion occurs when the joints bend and extension occurs when the joints straighten (**Figure 5.9D and E**). In the trunk, the vertebral column as a whole, flexion refers to bending the trunk forward and extension refers to the reverse movement. Lateral flexion of the trunk occurs when the trunk bends to the side about an anteroposterior axis.

Some bones, such as the humerus at the shoulder and the femur at the hip, can rotate axially, i.e. rotate about an axis along, parallel to or close to parallel to the long axis of the moving bone. For example, internal rotation (medial rotation) and external rotation (lateral rotation) of the shoulder occurs when the palm of the hand (relative to the anatomical position and not involving supination and pronation of the forearm) is rotated towards and away from the median plane.

In addition to abduction, adduction, flexion, extension and axial rotation, there are a number of other terms used to describe specific movements in certain joints. These movements include supination and pronation of the forearm and supination and pronation of the foot (see Chapter 7).

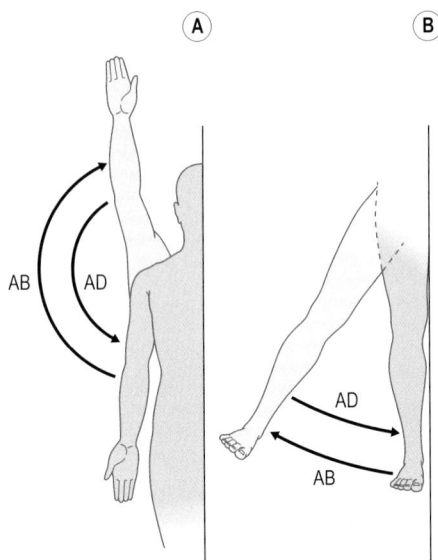

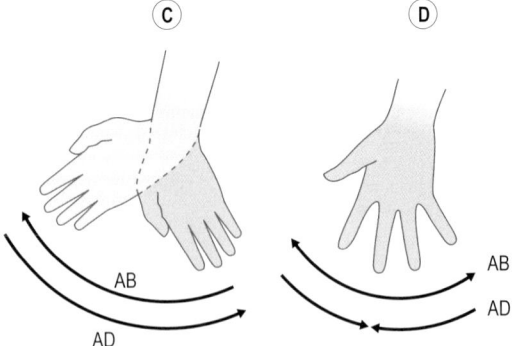

Figure 5.8 Abduction and adduction of (A) the shoulder joint, (B) the hip joint, (C) the wrist joint, and (D) the fingers.

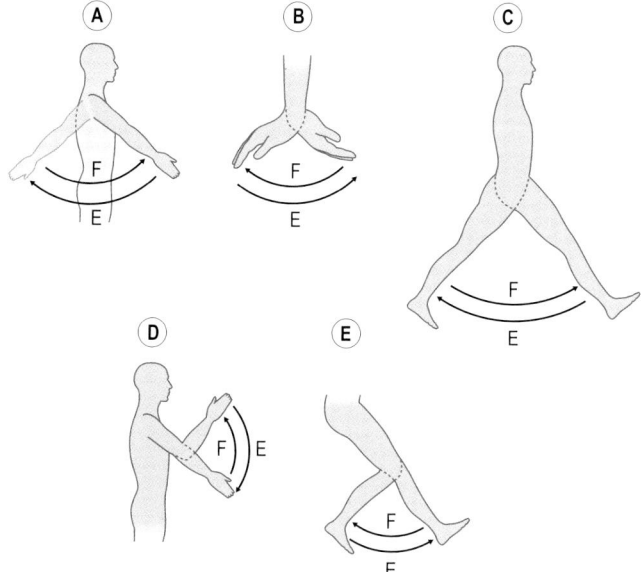

Figure 5.9 Flexion and extension of (A) shoulder joint, (B) wrist joint, (C) hip joint, (D) elbow joint and (E) tibiofemoral joint.

Synovial joint classification

Synovial joints are classified according to the type of movement that occurs in the joints. There are two kinds of synovial joint:

1. Joints in which the main type of movement is linear. Sliding occurs in all synovial joints to a certain extent, but in these joints, called gliding or plane joints, the articular surfaces are normally fairly flat and slide on each other in one or more directions. Gliding joints include the intercarpal and intertarsal joints and the joints between the superior and inferior facets of the vertebrae.

2. Joints in which the main type of movement is angular. Movement in these joints is normally a combination of rolling and sliding between the articular surfaces. There are three groups: uniaxial, biaxial and multiaxial.

Uniaxial

In uniaxial joints, movement normally takes place mainly about a single axis. There are two types of uniaxial joint, hinge joints and pivot joints.

In a hinge joint, a convex, pulley-shaped (bicondylar) articular surface articulates with a reciprocally shaped concave surface. The elbow (humeroulnar), interphalangeal and ankle joints are hinge joints (Figure 5.10). The notch of the pulley prevents (or severely limits) side-to-side movement. The tibiofemoral joint is usually regarded as a hinge joint even though the articular surfaces of the femoral and tibial condyles are not very congruent. However, in the normal tibiofemoral joint, the congruence between the articular surfaces is considerably increased by the presence of menisci.

In a pivot joint, a cylindrical articular surface rotates about its long axis within a ring formed by bone and fibrous tissue. The proximal radioulnar joint is a pivot joint (Figure 5.11). The head of the radius is held against the radial notch by a ligament called the annular ligament (annulus = ring). During supination and pronation of the forearm, the head of the radius rotates within the ring formed by the annular ligament and radial notch.

Biaxial

In biaxial joints, movement takes place mainly about two axes at right angles to each other, usually the anteroposterior (abduction/adduction) and mediolateral (flexion/extension) axes. There are three types of biaxial joint: condyloid, ellipsoid and saddle. In a condyloid joint a convex condylar surface articulates with a concave condylar surface. The metacarpophalangeal joints are condyloid joints. In an ellipsoid joint such as the radiocarpal joint, an elliptical convex surface articulates with an elliptical concave surface. The articular surface on the distal end of the radius is elliptical, concave and shallow. This surface articulates with the proximal articular surfaces of the scaphoid and lunate that together form a convex elliptical articular surface. Movements at the metacarpophalangeal joints and radiocarpal joints are normally combinations of flexion, extension, abduction and adduction. In a saddle (or sellar) joint, the articular surfaces are saddle shaped. Each articular surface is convex in one direction and concave in a direction at right angles to the convex direction. Movement takes place mainly in two planes at right angles to each other. The carpometacarpal joint of the thumb and the calcaneocuboid joints are saddle joints.

Multiaxial

Some joints, such as the shoulder and hip, can rotate about all three reference axes. By combining rotations about the three reference axes, these joints can rotate about any axis in between the three. Consequently,

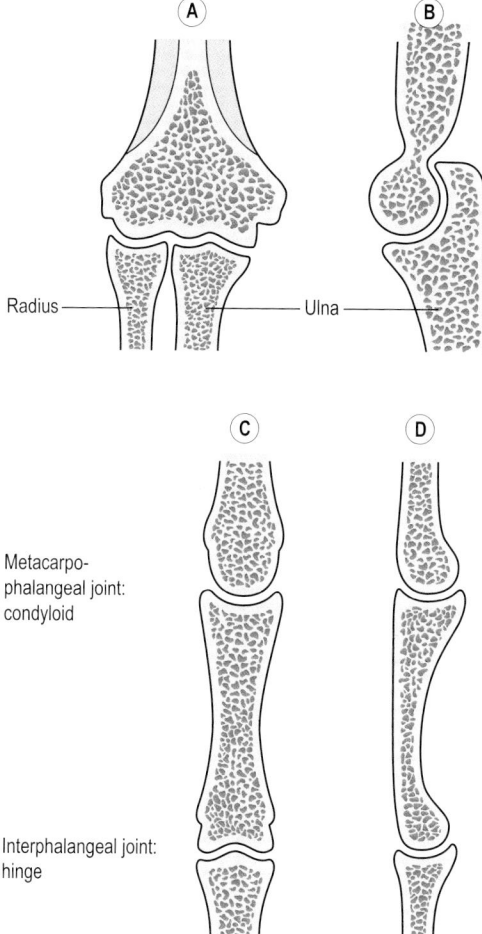

Radius —————   ————— Ulna —————

Metacarpo-
phalangeal joint:
condyloid

Interphalangeal joint:
hinge

Figure 5.10 Hinge and condyloid joints. (A) Coronal section through the right elbow joint in extension. (B) Sagittal section through the right humeroulnar joint in extension. (C and D) Coronal and sagittal sections, respectively, through metacarpophalangeal and interphalangeal joints in extension.

these joints are referred to as multiaxial joints. In this type of joint, a hemispherical articular surface articulates with a cup-like concavity. Owing to the shapes of the articular surfaces, these joints are usually referred to as ball and socket joints.

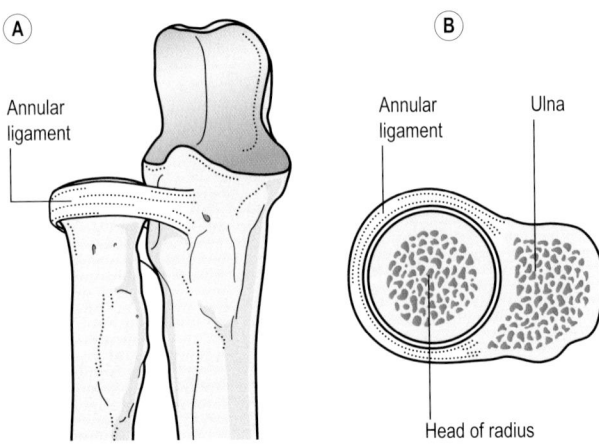

Figure 5.11 A typical pivot joint. (A) Anterior aspect of the right proximal radioulnar joint. (B) Transverse section through the right proximal radioulnar joint.

> **Key** *Concepts*
>
> Synovial joints are classified according to the type of movements that occur in the joints. Joints in which the main type of movement is linear are called gliding or plane joints. Joints in which the main type of movement is angular are classified into uniaxial (hinge and pivot), biaxial (condyloid, ellipsoid and saddle) and multiaxial (ball and socket) joints

Flexibility, stability and laxity in synovial joints

In previous sections of this chapter the terms flexibility and stability have been used in a general sense. At this point it is necessary to define these terms more specifically in relation to synovial joints.

Flexibility

In a synovial joint, flexibility refers to the range of movement in those directions (degrees of freedom) considered normal for the joint. For example, the tibiofemoral joint is designed primarily to rotate about a mediolateral axis, i.e. flexion and extension. Consequently, flexion and extension are considered normal movements at the tibiofemoral joint.

However, rotations about an anteroposterior axis, abduction and adduction, are considered abnormal movements at the tibiofemoral joint. The flexibility of a joint is determined by four factors:

1. Shape of the articular surfaces.
2. Tension in the joint capsule and ligaments at the ends of the various ranges of motion.
3. Soft tissue bulk, mainly skeletal muscle, surrounding the bones forming the joint.
4. Extensibility of the skeletal muscles controlling the movement of the joint.

Shape of articular surfaces

In some joints, flexibility is limited by the impingement or interlocking of the non-articular surfaces of the bones at the end of certain ranges of motion. For example, elbow extension is restricted by the interlocking of the olecranon process of the ulna with the olecranon fossa of the humerus (Figure 5.12A). Similarly, elbow flexion is restricted by the interlocking of the coronoid process of the ulna with the coronoid fossa of the humerus (see Figure 5.12B). Lateral and medial displacement of the radius and ulna relative to the humerus in the elbow joint is severely restricted by the interlocking of the articular surfaces (see Figure 5.12C).

Tension in joint capsule and ligaments

The function of ligaments is described in detail later in the chapter. At this point it is sufficient to appreciate that in a normal joint, the joint capsule and some of the ligaments that support the joint become taut at the end of each range of movement, thereby restricting further movement (see Figure 5.12).

Soft tissue bulk

Soft tissue bulk restricts flexibility in some joints. For example, elbow and knee flexion may be restricted in a heavily muscled individual due to the impingement of the adjacent body segments.

Extensibility of muscles

Muscle extensibility is the maximum length that a muscle–tendon unit can attain without injury. Muscle extensibility largely depends on the length range in which the muscle normally functions, i.e. the difference between its length when shortened and its length when extended. Generally speaking, the shorter the length range, the lower the extensibility. In the absence of regular flexibility training, games and sports involving highly repetitive and exclusive movement patterns are likely to result in reduced

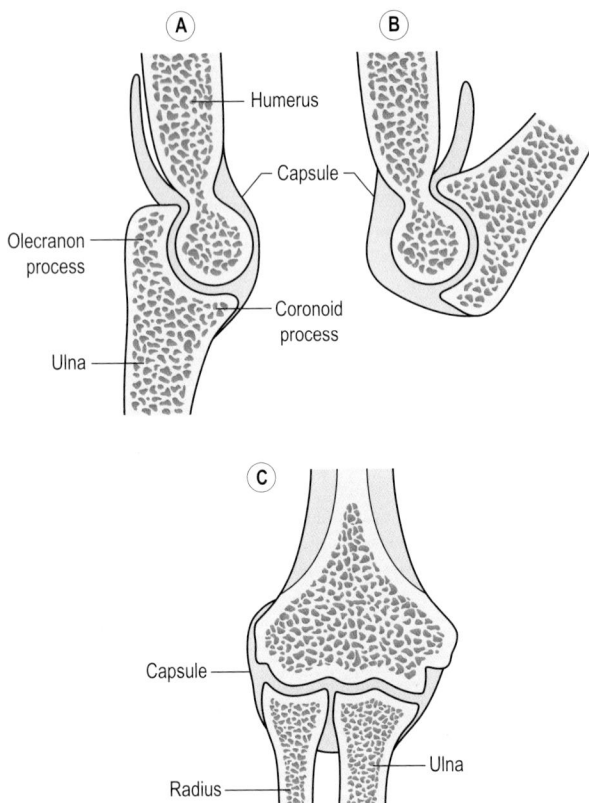

Figure 5.12 Effect of capsule and shape of articular surfaces on extension–flexion range of motion and abduction–adduction range of motion in the elbow joint. (A) Sagittal section through the elbow joint in extension. (B) Sagittal section through the elbow joint in flexion. (C) Coronal section through the elbow joint in extension.

extensibility in some muscles. In most individuals, muscle extensibility is probably the main factor limiting joint flexibility (Nicholas & Marino 1987; Herbert 1988).

Stability and laxity

In a synovial joint, stability refers to the degree of congruence between the articular surfaces. During joint movement, different parts of the articular surfaces come into contact with each other. However, in all positions of a joint, the better the congruence, the more stable the joint and the lower the risk of abnormal joint movements. The term laxity refers to the degree of instability in a joint, i.e. the range of movement in those directions considered abnormal for the joint. Joint congruence is determined by three factors:

1. Shape of the articular surfaces.
2. Suction between the articular surfaces.
3. Supporting structures: skeletal muscles, ligaments and the joint capsule.

Shape of articular surfaces

Most synovial joints have fairly congruent articular surfaces in all joint positions. In those joints where the articular surfaces are not very congruent, articular discs or menisci are usually present to improve congruence. Perfect congruence between articular surfaces would maximize joint stability and minimize joint laxity. However, perfect congruence would impair nutrition of the articular cartilage by restricting the flow of synovial fluid over the articular surfaces. Consequently, most joints are slightly incongruent and usually have a slight degree of laxity.

Suction between articular surfaces

In a normal joint, the articular surfaces are usually in close contact, with a thin film of synovial fluid between the surfaces. The film of fluid allows sliding between the surfaces but keeps the surfaces in contact with each other due to suction. In a similar manner, it is often difficult to separate two sheets of glass held in contact with each other by a thin film of water, even though the two sheets will slide on each other.

Supporting structures

The supporting structures largely responsible for maintaining close contact between the articular surfaces in all positions of a joint are the skeletal muscles that control the movement of the joint and the ligaments of the joint.

Functions of joint capsule and ligaments

The joint capsule has two main functions:

1. To assist joint stabilization by helping prevent (a) movement beyond normal range(s) and (b) excessive laxity.
2. To provide a base for the synovial membrane.

At the end of each normal range of movement, part of the joint capsule will usually become taut to prevent movement beyond the normal range. For example, in full extension of the elbow, the anterior aspect of the capsule will be taut and the posterior aspect will be slack (see **Figure 5.12A**). Similarly, in full flexion of the elbow, the posterior aspect of the capsule will be taut and the anterior will be slack (see **Figure 5.12B**). With regard to abnormal ranges of movement, the regions of the joint capsule in the planes of abnormal ranges of movement are usually at their natural length, i.e. neither taut nor slack. However, these regions quickly become taut in response to abnormal movements, thereby helping to restrict the ranges of abnormal movements. For example, abduction and adduction are abnormal movements at the elbow and the joint capsule normally helps to prevent these movements (see **Figure 5.12C**).

In association with skeletal muscles, ligaments bring about normal movement in joints. In a normal joint, ligaments help to maintain maximum stability in all positions of the joint by guiding the movements of the joint. During normal movements the tension in ligaments is low to moderate; ligaments only become taut to prevent or restrict abnormal movements. An abnormal movement may be defined as any movement that results in a decrease in normal congruence, i.e. any movement that results in partial or complete distraction (separation) of the articular surfaces. Partial distraction, when the area of contact between the articular surfaces is reduced, is called subluxation (sub = towards, luxation = dislocation).

Subluxations are usually transient; normal congruence is usually restored as soon as the load causing the subluxation is removed. Subluxations often result in joint sprains, i.e. partial tearing of ligaments and the joint capsule, together with effusion (swelling) in the joint. Complete distraction, when the articular surfaces are completely separated, is called luxation. Like subluxations, luxations are usually transient. A complete distraction that persists after the load causing it is removed is a dislocation (**Figure 5.13**). A dislocation usually requires manipulation by a physician or paramedic to restore the normal relationship between the articular surfaces (Grana et al 1987). Dislocation usually results in considerable damage to ligaments and the joint capsule.

Considerable force is required to cause severe subluxations and luxations. Ligaments and joint capsules are likely to be subjected to very high forces in two particular situations:

1. Unexpected situations in which the degree of muscular control of joint movement is less than adequate, for example, twisting an ankle by stepping on an uneven surface. In this situation, with the body moving forward over the foot, the ankle is likely to be rapidly and forcibly twisted.
2. High-speed collisions, such as a tackle in football or soccer.

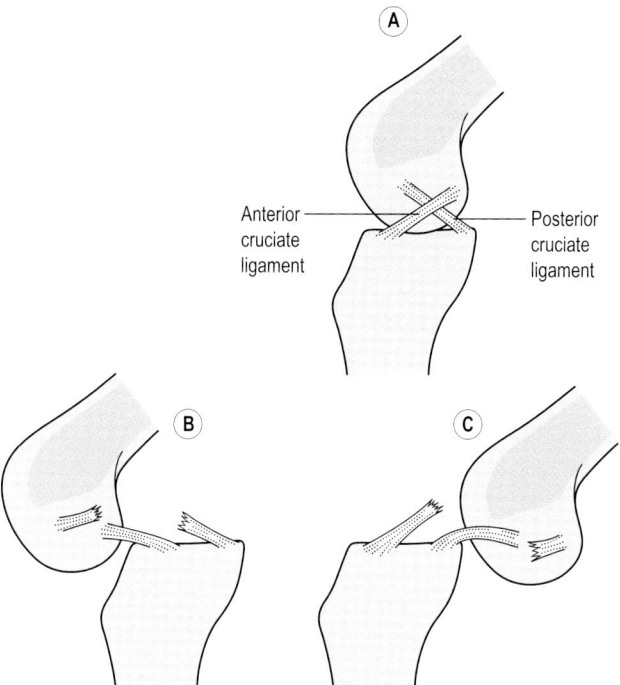

Figure 5.13 Dislocation of the tibiofemoral joint. (A) Normal orientation of the cruciate ligaments in the partially flexed tibiofemoral joint. (B) Anterior dislocation of the femur on the tibia showing torn posterior cruciate ligament. (C) Posterior dislocation of the femur on the tibia showing torn anterior cruciate ligament.

There are basically two kinds of abnormal movements in joints: hyperflexibility and excessive laxity. These will be described with reference to the tibiofemoral joint.

Hyperflexibility

Hyperflexibility may be defined as movement beyond the normal range of movement in a direction considered normal for the joint. Extension is a normal movement in the tibiofemoral joint. At full extension of the tibiofemoral joint, the four main ligaments that support the joint will become taut to prevent further movement. If the joint is forcibly extended beyond this position, i.e. hyperextended, the ligaments and joint capsule will be damaged (**Figure 5.14**). Hyperextension of the tibiofemoral joint is an

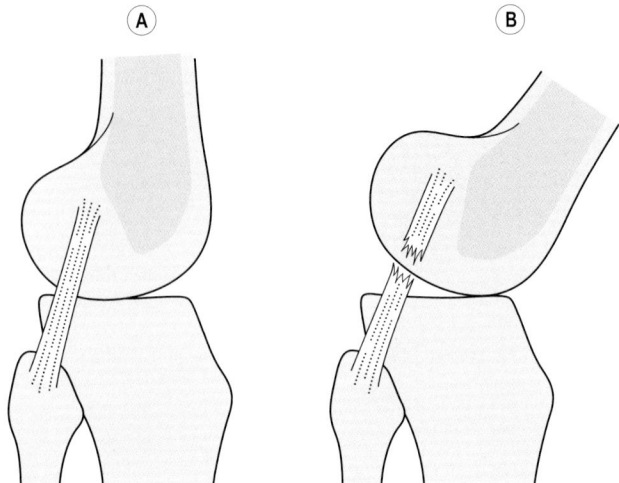

Figure 5.14 Hyperextension of the tibiofemoral joint. (A) Lateral aspect of the extended right tibiofemoral joint showing the normal orientation of the lateral ligament. (B) Hyperextension of the tibiofemoral joint showing torn lateral ligament.

example of hyperflexibility. In most individuals, full extension of the tibiofemoral joint normally corresponds to a position in which the leg is straight, i.e. the upper and lower leg are in line. Consequently, in most individuals, hyperextension of the tibiofemoral joint occurs when the joint is extended beyond the straight position. However, some individuals are able to voluntarily extend their tibiofemoral joints slightly beyond the straight position. In these cases, extension of the joints beyond the straight position would be considered normal for the individual provided that normal congruence was maintained.

Excessive laxity

Excessive laxity is defined as movement beyond the normal degree of laxity in directions considered abnormal for the joint. For example, abduction and adduction are abnormal movements in the tibiofemoral joint. In a normal tibiofemoral joint these movements are restricted to a minimal level by the medial and lateral ligaments, respectively (Figure 5.15). Abduction of the tibiofemoral joint beyond a minimal level will damage the medial ligament and joint capsule. Similarly, adduction of the joint beyond a minimal level will damage the lateral ligament and joint capsule.

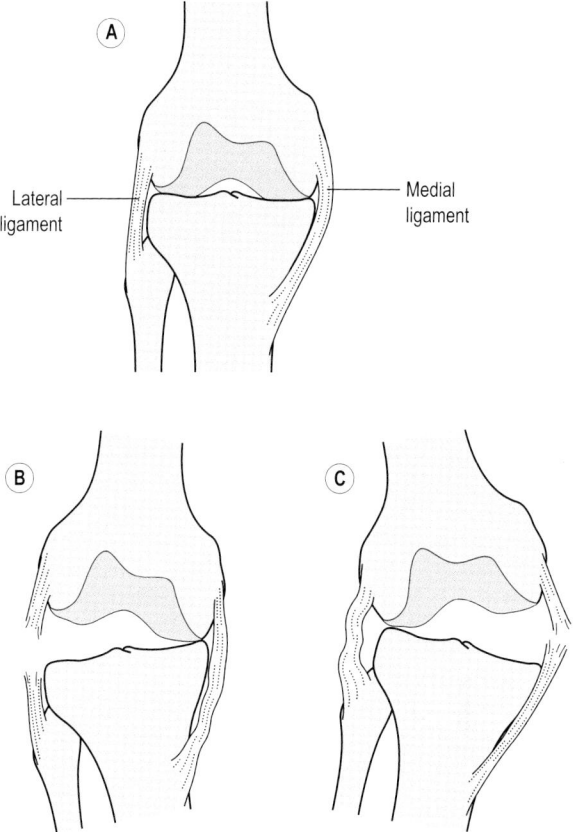

Figure 5.15 (A) Anterior aspect of the right tibiofemoral joint showing the normal orientation of the lateral and medial ligaments. (B) Adduction of the tibiofemoral joint showing torn lateral ligament. (C) Abduction of the tibiofemoral joint showing torn medial ligament.

Movements of the femoral condyles backward and forward (along the anteroposterior axis) relative to the tibial condyles are abnormal movements. In a normal tibiofemoral joint, these movements are restricted to a minimal level by the anterior and posterior cruciate ligaments, respectively (see **Figure 5.13**). Forward movement of the femoral condyles beyond a minimal level will damage the posterior cruciate ligament and the joint capsule. Similarly, backward movement of the femoral condyles beyond a minimal level will damage the anterior cruciate ligament and the joint capsule.

Any degree of hyperflexibility or excessive laxity in a joint tends to damage not only ligaments and joint capsules, but also the articular cartilage. The damage to articular cartilage is due to localized overloading as a result the abnormal movements.

Movement between articular surfaces

There are three basic types of movement that occur between articular surfaces: spinning, sliding and rolling. In most joints, normal movement involves a combination of these three types of movement, either simultaneously or sequentially. Sliding, and to a lesser extent spinning, subjects the articular cartilage to compression and shear loading. Rolling subjects the articular cartilage to compression loading. These loads tend to cause mechanical wear of the articular cartilage in the form of damage to the surface of the cartilage and to the internal structure of the cartilage. Under normal circumstances, when the articular surfaces are congruent and properly lubricated, there is a balance between the amount of wear and the production of new cartilage. However, during subluxations, those parts of the articular cartilage that remain in contact are subjected to abnormally high loads. Over a period of time, the wear caused by the abnormal loads may outpace the production of new cartilage and result in permanent damage to the cartilage.

Review questions

1. Describe the two main functions of joints.
2. Describe the two main types of joint in terms of structure.
3. Differentiate between a syndesmosis and a symphysis.
4. Describe the basic structure of a synovial joint.
5. With respect to synovial joints, differentiate between the following:
   - Flexibility and laxity
   - Stability and congruence
   - Subluxation and dislocation
6. Describe the factors affecting the flexibility of a synovial joint.
7. Describe the function of ligaments in relation to synovial joints.
8. Describe the types of movement that occur between the articular surfaces in synovial joints.

References

Basmajian JV (1970) Primary anatomy. Williams & Wilkins, Baltimore.

Grana WA, Holder S, Schelberg-Karnes E (1987) How I manage acute shoulder dislocations. The Physician and Sportsmedicine 15:88–93.

Herbert R (1988) The passive mechanical properties of muscles and their adaptations to altered patterns of use. Australian Journal of Physiotherapy 34:141–49.

Nicholas JA, Marino M (1987) The relationship of injuries of the leg, foot and ankle to proximal thigh strength in athletes. Foot & Ankle 7:218–28.

The neuromuscular system

The muscular system is the interface between the nervous and skeletal systems. The muscles produce the forces to move the levers of the skeletal system, but the nervous system determines the level and timing of the muscle forces. The nervous system constantly monitors and interprets information from the various senses concerning body position and body movement, including information from muscles and joint-supporting structures, and on the basis of this information sends instructions to the muscles to coordinate body movement in both voluntary and involuntary (reflex) movements. Those parts of the nervous and muscular systems responsible for bringing about coordinated movement are referred to as the neuromuscular system. This chapter describes the structure and function of the neuromuscular system.

The nervous system

The nervous system consists of approximately 13 000 million nerve cells called neurones and an equally large number of specialized connective tissue cells called glial cells. Neurones are specialized to conduct electrochemical impulses rapidly throughout the body to coordinate all the essential biological functions. The cells of the nervous system are organized into two functional divisions and two structural divisions (Williams et al 1995). The functional divisions are the cerebrospinal nervous system and the autonomic nervous system.

The cerebrospinal nervous system, also known as the somatic, craniospinal or voluntary nervous system, is under voluntary control except for reflex movements. A reflex movement provides protection by rapidly removing part of the body from a source of danger without conscious effort. The cerebrospinal nervous system includes those parts of the nervous system concerned with consciousness and mental activities, and control of skeletal muscle. The autonomic nervous system, also known as the visceral or involuntary nervous system, is not under voluntary control; it includes those parts of the nervous system that control the visceral muscles, the heart, and the exocrine and endocrine glands.

The two structural divisions of the nervous system are the central nervous system and the peripheral nervous system. The central nervous system consists of the brain and spinal cord. The peripheral nervous system consists of 43 pairs of nerves (bundles of nerve fibres), which arise from the base of the brain and the spinal cord (Figure 6.1). The upper 12 pairs of nerves arise from the base of the brain and are called cranial nerves. The other 31 pairs arise from the spinal cord and are called spinal nerves. The cranial and spinal nerves convey information between the central nervous system and the rest of the body.

Neurones

Neurones differ in size and shape, but they all have three common structural features: a cell body, processes of varying length that extend from the cell body, and specialized sites for communicating with other neurons, with specialized receptors such as pain receptors, and with specialized effectors such as motor end plates in muscle. Neurones are classified by the direction in which they conduct impulses in relation to the brain. Sensory or afferent neurones conduct impulses towards the brain, and motor or efferent neurones conduct impulses away from the brain.

Nerve fibres

The processes that extend from the cell bodies of neurones are called nerve fibres. Nerve fibres vary in length from a few millimetres to more than 1 m. There are two types of nerve fibre, dendrites and axons. Dendrites, or afferent fibres, conduct impulses towards the cell body. Axons, or efferent fibres, conduct impulses away from the cell body. In addition to the sensory and motor classification, neurones are classified on the basis of the number of processes arising from the cell body into pseudounipolar, bipolar and multipolar.

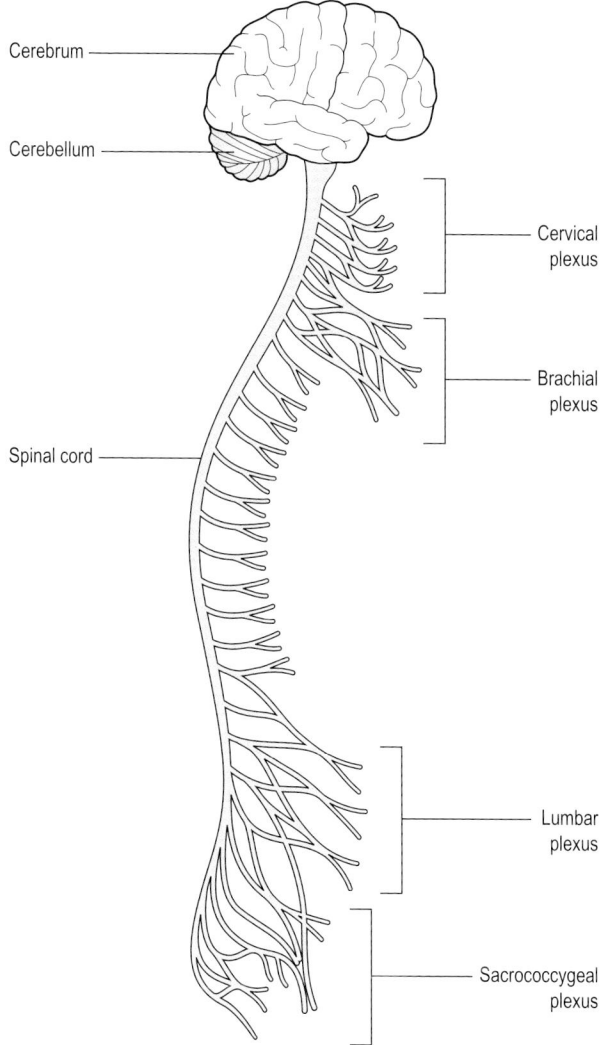

Figure 6.1 The central nervous system consists of the brain (cerebrum and cerebellum) and spinal cord. The peripheral nervous system consists of the spinal nerves. Many of the spinal nerves link together to form plexuses. All of the spinal nerves and plexuses branch profusely throughout the regions of the body that they innervate.

In a pseudounipolar neurone, there appears to be one process arising from the cell body that quickly divides into an afferent fibre and an efferent fibre (Figure 6.2A). Sensory neurones in the peripheral nervous system are pseudounipolar neurones. A bipolar neurone has two distinct processes, one afferent and one efferent (see Figure 6.2B). Bipolar neurones are found in the sensory areas of the eye, ear and nose. Multipolar neurones have numerous relatively short dendrites with a single axon that may branch at various points (see Figure 6.2C). Most of the neurones in the brain and spinal cord are multipolar neurones.

Myelinated and non-myelinated nerve fibres

Glial cells provide mechanical and metabolic support to neurones. In the central nervous system there are a variety of glial cells including astrocytes, which provide support for blood vessels, and oligodendrocytes, which provide support for nerve fibres. In the peripheral nervous system, there is only one type of glial cell, Schwann cells (Gamble 1988). All nerve fibres of the peripheral nervous system are enveloped by Schwann

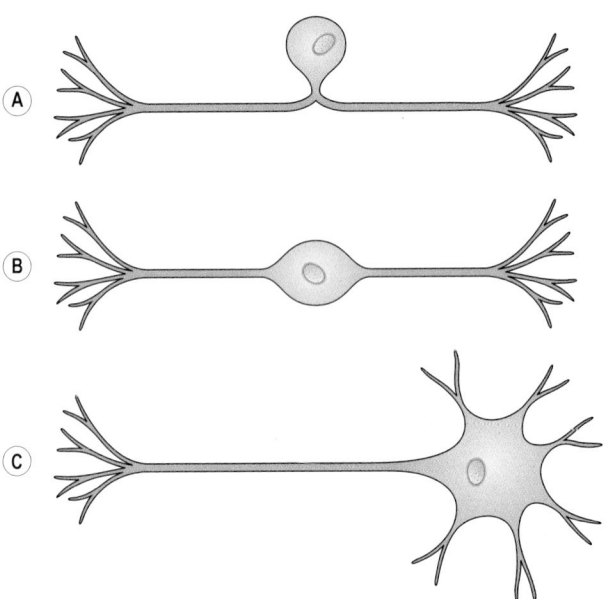

Figure 6.2 Types of neurones. (A) Pseudounipolar. (B) Bipolar. (C) Multipolar.

cells, which provide the same type of mechanical support as the oligo-dendrocytes provide for nerve fibres in the central nervous system. The Schwann cells around some fibres produce a fatty substance called myelin, which is deposited around the fibres as a multilayered myelin sheath. The sheath is in the form of a spiral with up to 100 regularly spaced layers of myelin separated by folds of Schwann cell membrane. The outer fold of the Schwann cell membrane is referred to as the neuri-lemma (Figure 6.3).

The greater the number of layers of myelin in the sheath, the faster the speed of nerve transmission along the nerve fibre. Nerve fibres that have a myelin sheath are referred to as myelinated or medullated nerve fibres and those nerve fibres that do not are referred to as non-myelinated or non-medullated nerve fibres. Each myelinated nerve fibre is enclosed within a continuous chain of Schwann cells and each Schwann cell envelops approximately 1 mm of nerve fibre. At the junction between

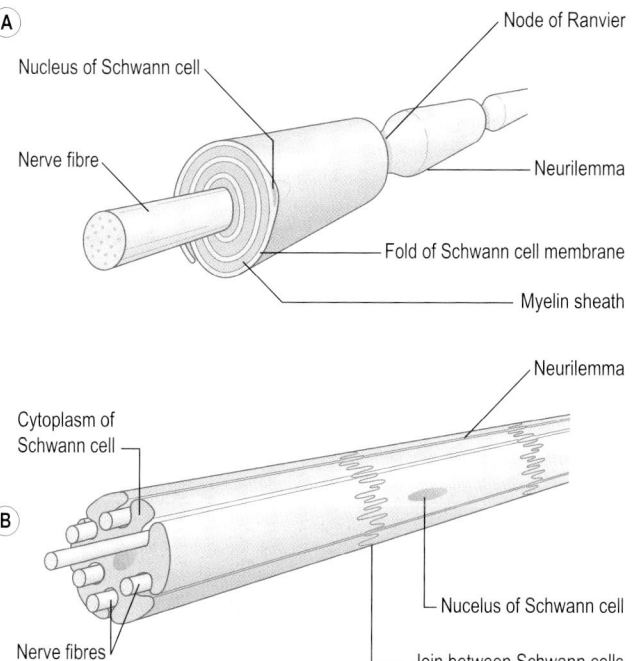

Figure 6.3 Nerve fibres. (A) Myelinated. (B) Non-myelinated.

adjacent Schwann cells the myelin sheath is interrupted such that the nerve fibre is only covered by the neurilemma; these regions are referred to as nodes of Ranvier (see **Figure 6.3A**). The nodes of Ranvier facilitate intercellular exchange between the nerve fibres and the surrounding extracellular fluid, which is important for the nutrition of the nerve fibre and for the transmission of impulses along the nerve fibre. Myelinated nerve fibres in the central nervous system are different from those in the peripheral nervous system in that they are not surrounded by Schwann cells (there are no Schwann cells in the central nervous system) and have no neurilemma. It is thought that oligodendrocytes are responsible for the formation of the myelin sheath around these fibres at the embryonic stage (Williams et al 1995).

Non-myelinated nerve fibres of the peripheral nervous system are also enclosed within continuous chains of Schwann cells. However, in comparison to myelinated fibres, there are no nodes of Ranvier in non-myelinated fibres and there may be as many as nine nerve fibres enveloped within the folds of the Schwann cells in the chain (see **Figure 6.3B**).

Nerve fibre endings

The dendrites and axons of all types of neurones have a large number of terminal branches or nerve endings devoid of Schwann cells and myelin sheath. There are three types of nerve endings:

1. *Sensory*: where the nerve ending is in contact with a specialized receptor organ such as a pain receptor.

2. *Motor*: where the nerve ending is in contact with a specialized effector organ such as a motor end plate in muscle.

3. *Synapse*: where the nerve ending is in contact with another neurone.

Sensory and motor nerve endings are referred to as end organs. Neurones that only have synapses at their nerve endings are called association neurones or interneurones.

Nerve impulse transmission

The cytoplasm of a neurone and the extracellular fluid surrounding the neurone contain many different ions (electrically charged atoms). These include positively charged inorganic ions such as sodium (Na^+) and potassium (K^+), negatively charged inorganic ions such as chloride (Cl^-) and various organic anions (negatively charged amino acids and proteins [A^-]).

Like most membranes in the body, a nerve fibre membrane is semi-permeable, i.e. it has a large number of tiny holes through which ions and small molecules (aggregations of ions) pass from one side of the membrane to the other. The movement of ions through the nerve fibre membrane depends on the permeability of the membrane (the number and size of the holes in the membrane) and the force tending to drive the ions through the membrane. The driving force has electrical, chemical and, in the case of Na^+ and K^+, mechanical components. The electrical component depends on the polarity of the ions; like charges repel each other and unlike charges attract each other. The chemical component depends on the concentration of ions in different regions; ions move from an area of high concentration to areas of lower concentration. The mechanical component results from specialized regions of the membrane collectively referred to as the Na^+-K^+ pump. The Na^+-K^+ pump transports Na^+ out of the cytoplasm and into the extracellular fluid and transports K^+ in the opposite direction.

Under resting conditions, when the nerve fibre is not transmitting an impulse, the net effect of the membrane permeability and the driving forces on the various ions is that the electrical charge on the outside of the fibre membrane is approximately 70 mV (millivolts) higher than on the inside; the potential difference across the membrane is approximately 70 mV, with the inside negative with respect to the outside (Figure 6.4A). This resting potential difference is referred to as resting membrane potential (RMP) (Enoka 1994).

The arrival of a stimulus at a nerve fibre in a state of RMP alters the permeability of the fibre membrane to Na^+ and K^+ so that Na^+ flows into the cell and K^+ flows out of the cell. The flow of Na^+ into the cell is initially greater than the flow of K^+ out of the cell so that the potential difference across the membrane decreases. If the decrease in potential difference, which depends on the strength of the stimulus, reaches a critical level of approximately 60 mV (with the inside of the membrane negative with respect to the outside), the membrane will be depolarized, i.e. the potential difference across the fibre membrane rapidly changes by approximately 100 mV from 70 mV, with the inside of the membrane negative with respect to the outside, to approximately 30 mV, with the inside of the membrane positive with respect to the outside (see Figure 6.4A).

This change in potential difference constitutes an action potential, which results in the flow of electrical current, called local current, between the depolarized region of the cell membrane and the adjacent unpolarized regions (both sides) (see Figure 6.4B and C). The establishment of local current results in progressive (rapid wave) depolarization of the rest of the cell membrane so that the impulse is transmitted along the whole length of the fibre. After depolarization, the membrane is rapidly depolarized such that the action potential appears as a spike in a graph of the change in membrane potential with time (see Figure 6.4A).

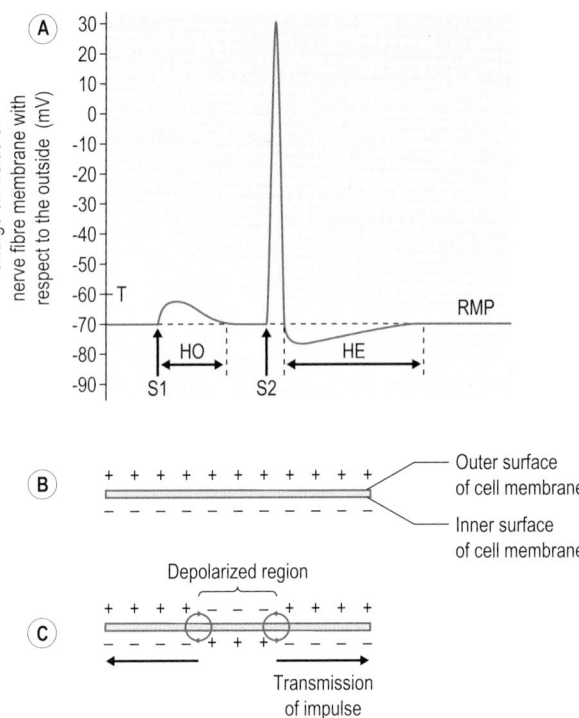

Figure 6.4 Resting membrane potential (RMP) and change in membrane potential. (A) Change in membrane potential in response to stimulation: T, level of hypopolarization necessary to trigger depolarization (T ≈ −60 mV); S_1, a stimulus that results in hypopolarization but not depolarization; H_O, period of hypopolarization following S_1, S_2, a stimulus that results in depolarization; H_E, period of hyperpolarization following repolarization. (B) Electrical charge on nerve fibre membrane during RMP. The charge on the inside of the membrane is approximately −70 mV with respect to the outside. (C) Electrical charge on depolarized region of nerve fibre membrane; the charge on the inside of the membrane is approximately +30 mV with respect to the outside.

The duration of the action potential spike (depolarization and repolarization) is less than one millisecond (Gamble 1988). Repolarization is due largely to a rapid decrease in the flow of Na^+ into the cell (due to reduced permeability of the membrane to Na^+ ions) and continued flow of K^+ out of the cell. Following repolarization there is usually a period (15–100 ms) of hyperpolarization in which the potential difference across the membrane is slightly greater than RMP as RMP is gradually restored (see **Figure 6.4A**).

Table 6.1 Classification of peripheral nerve fibres

Class	Speed (m/s)	Innervation
Afferent (sensory) fibres		
Ia	65–130	Muscle spindle intrafusal fibres
Ib	65–130	Golgi tendon organs
II	20–90	Muscle spindle intrafusal fibres, pressure receptors
III	12–45	Temperature and pain receptors
IV	0.2–2.0	Viscera, pain receptors
Efferent (motor) fibres		
Aα	65–130	Fast twitch extrafusal muscle fibres
Aβ	40–80	Slow twitch extrafusal muscle fibres, muscle spindle intrafusal fibres
Aγ	10–50	Muscle spindle intrafusal fibres
B	4–25	Presynaptic autonomics
C	0.2–2.0	Postsynaptic autonomics

Data from Gamble 1988.

A stimulus not strong enough to cause depolarization results in a period of hypopolarization prior to restoration of RMP, i.e. a period in which the potential difference across the membrane is slightly less than RMP (see Figure 6.4A).

Synapses

Every branch of an axon terminates in an end bulb (or end foot) that rests on the surface of a neighboring neurone to form a synapse, a specialized region that facilitates one-way communication between the two neurones. Each neurone synapses with hundreds or thousands of other neurones (Williams et al 1995). The most common type of synapse is axodendritic (between an axon and a dendrite), but synapses may also be axosomatic (between an axon and a cell body), and axoaxonic (between two axons).

Speed of nerve impulse transmission

The speed of transmission or conduction of impulses is directly proportional to fibre diameter and the thickness of the myelin sheath; the larger the diameter and the thicker the myelin sheath, the faster the speed of conduction. Non-myelinated fibres do not have a myelin sheath and, thus, have much lower conduction speeds than myelinated fibres. Nerve fibres of the peripheral nervous system are classified on the basis of conduction speed and fibre diameter (Table 6.1). There are five main categories of afferent fibres (Ia, Ib, II, III and IV) and five main categories of efferent fibres (Aα, Aβ, Aγ, B and C). Skeletal muscle is innervated by the fastest afferent fibres (Ia and Ib) and the fastest efferent fibres (Aα, Aβ, Aγ).

Nerve tissue organization in the brain

The brain is the largest and most complex aggregation of neurones in the nervous system. It consists of the cerebrum and the cerebellum (see Figure 6.1). The cerebrum occupies most of the cranium and is bigger than the cerebellum. The region of the cerebrum close to and including its surface is called the cerebral cortex and consists of grey matter, i.e. the cell bodies of neurones and their processes, which are largely non-myelinated, together with their synapses and supporting glial cells. The cerebrum is heavily convoluted with fissures of varying depth. The largest fissure is the longitudinal central fissure that divides the cerebrum in the median plane into right and left cerebral hemispheres.

The cerebellum, which occupies the posterior inferior aspect of the cranium, is separated from the cerebrum by the transverse fissure. Like the cerebrum, the region of the cerebellum close to and including its surface consists of grey matter and is called the cerebellar cortex. The cerebellum is not convoluted but is traversed by numerous small furrows. The convolutions of the cerebrum and furrows of the cerebellum significantly increase their surface areas and, thus, the volume of grey matter. The inner parts of the cerebrum and cerebellum consist of white matter, i.e. largely myelinated nerve fibres organized into groups that link the different parts of the cerebrum and cerebellum with each other and with the spinal cord.

Key Concepts

The brain is the largest and most complex aggregation of neurones in the nervous system. It consists of the cerebrum and the cerebellum. The cerebral cortex and cerebellar cortex consist of grey matter. The inner parts of the cerebrum and cerebellum consist of white matter

Nerve tissue organization in the spinal cord and spinal nerves

In transverse section, the spinal cord is roughly oval shaped with an anterior median fissure and a posterior medial septum (Figure 6.5). The central area is dominated by a roughly H-shaped mass of grey matter, with the

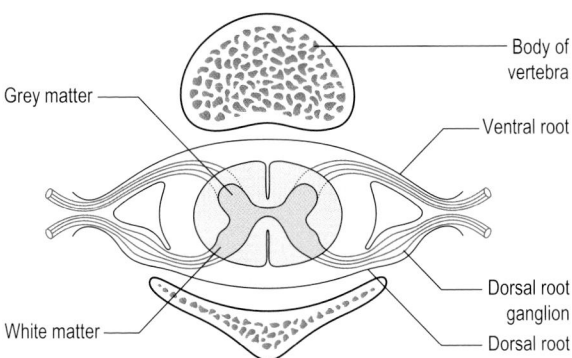

Figure 6.5 Transverse section through the spinal cord at the level of a pair of spinal nerves.

rest of the cord consisting of white matter in which the groups of fibres run parallel with the spinal cord. The posterior projections (or horns) of the grey matter are continuous with afferent fibres that enter the spinal cord via the left and right dorsal (posterior) roots of the corresponding left and right spinal nerves. The cell bodies of the afferent fibres are located in the dorsal root ganglia; a ganglion is an aggregation of cell bodies outside the spinal cord.

The anterior projections of the grey matter are continuous with efferent fibres that leave the spinal cord via the left and right ventral (anterior) roots of the corresponding spinal nerve. A spinal nerve is formed by the aggregation of the afferent and efferent fibres as they pass through the corresponding intervertebral foramen. Each spinal nerve divides into an anterior and posterior branch just outside the intervertebral foramen; the anterior branch is usually much larger than the posterior branch. In each branch, the individual nerve fibres are enveloped in a thin layer of connective tissue called the endoneurium, which supports a blood capillary network. The fibres are grouped together in bundles called fasciculi or funiculi and each fasciculus is sheathed within a layer of connective tissue called the perineurium. The fasciculi are grouped together and sheathed within another layer of connective tissue called the epineurium to form the complete spinal nerve (Figure 6.6). Each spinal nerve consists of a mixture of myelinated and non-myelinated nerve fibres.

In all regions of the spinal cord apart from most of the thoracic region, the spinal nerves link up with each other on each side of the spinal cord to form networks called plexuses (see Figure 6.1). The upper four pairs of cervical nerves form the left and right cervical plexuses that innervate the

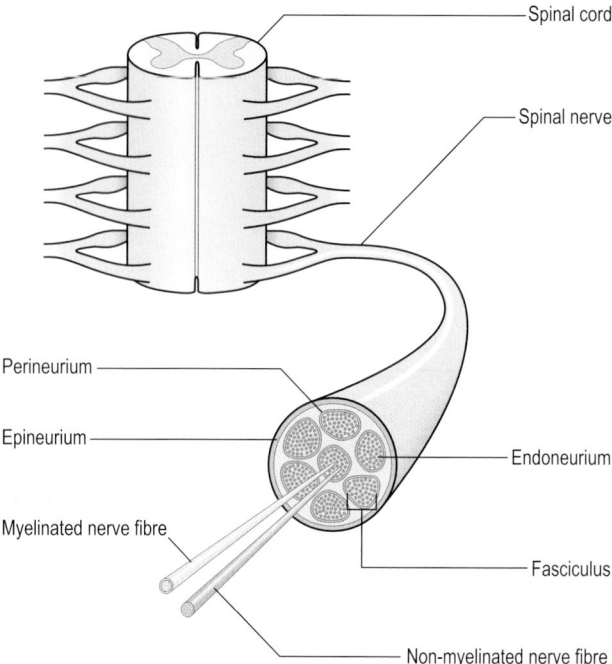

Figure 6.6 Structure of a spinal nerve and its relationship to the spinal cord.

head and upper part of the neck. The other cervical spinal nerves combine with the first pair of thoracic spinal nerves to form the left and right brachial plexuses that innervate the upper limbs. The twelfth thoracic spinal nerves combine with the lumbar spinal nerves to form the lumbar plexuses that innervate the lower trunk and pelvis. The sacral and coccygeal spinal nerves with branches from the fourth and fifth lumbar spinal nerves combine to form the left and right sacrococcygeal plexuses that innervate the legs. The second to the eleventh thoracic spinal nerves, which innervate the trunk, do not form plexuses.

Voluntary and reflex movements

In general terms, the brain interprets sensory information from the various receptors and on the basis of this information brings about appropriate responses via effectors to ensure normal bodily functioning. Whereas the autonomic nervous system operates largely at a subconscious level,

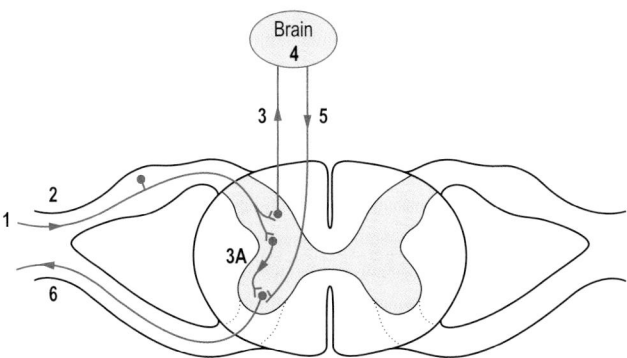

Figure 6.7 Pathway of impulses in voluntary and reflex movements (see text for description of stages).

the cerebrospinal nervous system normally operates at the level of consciousness, i.e. the individual consciously processes information, such as visual, auditory and kinaesthetic (sense of movement) information, and consciously brings about appropriate responses. For example, consider a child just about to touch a hot coffeepot resting on a stove (**Figure 6.7**).

1. As the child's hand moves close to the pot, the heat radiating from the pot may excite the heat receptors in the skin of her hand.
2. This information is transmitted by sensory neurones in the peripheral nervous system to the spinal cord, where the nerve endings of the axons of the sensory neurones synapse with many other neurones.
3. Some of these neurones relay the information to the brain centre responsible for heat sensation in the hand, and the child experiences the sensation of heat.
4. Having sensed the heat from the coffeepot, the child decides to move her hand away from the pot to avoid the danger.
5. Impulses are sent from the brain to motor neurones in the spinal cord that innervate the muscles of the arm.
6. The impulses are relayed to the muscles of the arm resulting in movement of the hand away from the coffeepot.

The above sequence, illustrated in **Figure 6.7**, is typical of all voluntary movements. If the child had not sensed the danger and had actually touched the hot coffeepot, she would have jerked her hand away from the pot with lightening speed before she was even conscious of the heat. This extremely rapid involuntary reaction is called a reflex action and is one of a host of similar reflexes that result in instant reactions to protect the body from potentially harmful stimuli. In this particular case, the reflex action results in the child's hand being in contact with the coffeepot for only a fraction of the time it would have taken for the child to take voluntarily (and, therefore, consciously) her hand away from the pot. Consequently, the child may sustain a relatively slight burn rather than a serious burn, which would have resulted from more prolonged contact with the coffeepot.

The increased speed of a reflex action compared to a voluntary action is due to a reduction in the distance over which the impulses travel from receptor to effector. In this particular case, the heat from the coffeepot is sensed by the heat receptors in the hand, which send impulses to the spinal cord. This is basically the same as in a voluntary action (stages 1 and 2 in **Figure 6.7**). However, in a reflex action the impulses are transmitted directly across the spinal cord to the motor neurones that innervate the muscles of the arm (stage 3A in **Figure 6.7**). The impulses are relayed to the muscles of the arm and the reflex action is completed. Consequently, a reflex action is faster than a voluntary action because no time is spent in transmitting impulses up the spinal cord to the brain, making a decision and transmitting impulses back down the spinal cord (stages 3, 4 and 5 in **Figure 6.7** are omitted). A reflex action is triggered when the intensity of the impulse output from the receptors is above a certain threshold indicating extreme danger to the body. In this particular case, the hotter the coffeepot, the greater the intensity of output from the heat receptors and the greater the likelihood of a reflex action being triggered.

As the impulses are transmitted across the spinal cord (stage 3A in **Figure 6.7**), the impulses are simultaneously transmitted to the brain as in a voluntary action (stage 3 in **Figure 6.7**). However, stage 3 (transmission of impulses to the brain and sensation of heat) takes longer than stages 3A and 6 such that it is a fraction of a second after the child jerks her hand away from the coffeepot before she perceives any pain in her hand.

Nerve fibre injuries

Injuries to peripheral nerves usually occur as a result of compression or traction (stretching) or a combination of the two (Kleinrensink et al 1994). Compression and traction may damage any or all of the main structures: connective tissue sheaths, Schwann cells, myelin sheath (when present) and nerve fibres. Compression and traction damage the nerve structures directly due to crushing and tearing, respectively. Prolonged compression also may damage the nerve structures indirectly as a result of ischemia, i.e. a disruption of the local blood supply (due to compression of the local capillary network), which results in a deficiency of oxygen and nutrients to the affected tissues.

Injuries to peripheral nerves are classified on the basis of degree of structural and functional damage into neuropraxia, axonotmesis and neurotmesis (Seddon 1972). Neuropraxia is the lowest level of damage. It involves damage to Schwann cells and myelin sheaths, but little or no damage to nerve fibres or endoneurium. Neuropraxia is characterized by a disruption to impulse transmission, which is often associated with pain and tingling in the areas innervated by the affected nerves. Recovery of a nerve fibre from neuropraxia usually occurs within 10 to 14 days. In axonotmesis, the nerve fibre and Schwann cell covering or myelin sheath are severed, but the endoneurium remains intact. The severed part of the fibre degenerates. Recovery begins when fibre sprouts emerge from the severed end of the part of the fibre still attached to the cell body. One of the sprouts eventually dominates and gradually grows along the tube formed by the endoneurium. Growth occurs at a rate of approximately 2.5 cm per month, and function gradually returns to normal as the fibre and myelin sheath (when present) returns to normal. In neurotmesis, the nerve fibre, Schwann cell covering or myelin sheath, and endoneurium are all severed. The severed part of the fibre degenerates and without surgical repair it is unlikely that much recovery will occur. With surgical repair some recovery occurs similarly to the way in which a fibre recovers from axonotmesis, but functional outcome is often less than satisfactory.

Key *Concepts*

> Compression and traction can damage nerves directly due to crushing and tearing, respectively. Prolonged compression may damage the nerve structures indirectly as a result of ischemia

Skeletal muscle structure

The composition and basic function of the muscular system are described in Chapter 2. This section describes the macrostructure of skeletal muscle.

Origins and insertions

Most of the muscles are attached to the skeletal system by tendons or aponeuroses. However, one or both attachments of some muscles attach directly onto bone without an intervening tendon or aponeurosis.

For example, the biceps femoris arises from two sites: a tendinous attachment to the ischial tuberosity and directly from a long narrow area on the lateral aspect of the femur (Figure 6.8A). The muscle is attached by a single tendon to the head of the fibula and adjoining posterior aspect of the lateral tibial condyle. Most of the muscles of the upper and lower limbs are arranged in line with the direction of the long bones. For descriptive purposes, the proximal and distal attachments of each of these muscles are referred to as the origin and insertion of the muscles, respectively. For example, the origin of the biceps femoris is on the ischial tuberosity (long head) and femur (short head) and the insertion is on the fibula and tibia (see Figure 6.8B). The origins and insertions of the muscles of the trunk

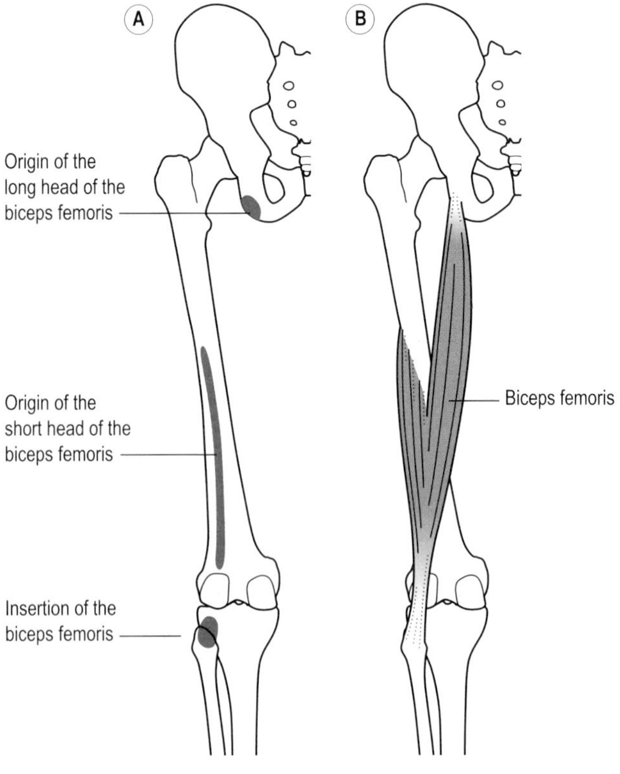

Origin of the
long head of the
biceps femoris —

Origin of the
short head of the
biceps femoris —

— Biceps femoris

Insertion of the
biceps femoris —

Figure 6.8 Attachments of the biceps femoris.

and the muscles that link the trunk to the limbs tend to be medial and lateral, respectively, but there are exceptions.

Pennate and non-pennate muscles

A skeletal muscle is made up of skeletal muscle cells bound by various layers of connective tissue that support extensive networks of nerves and blood vessels. Muscle cells are long and thin and, as such, they are usually referred to as muscle fibres. Each muscle fibre is approximately 50μm (1 μm = one-millionth of a metre) wide. All the fibres in an individual muscle are about the same length. In some muscles the fibres are relatively short; for example, in the muscles that move the eyes, the fibres are 2 to 4 mm long. In contrast, the fibres in some other muscles are very long; for example, the fibres of the sartorius are approximately 30 cm long. The muscle fibre length in most muscles is between these two extreme values.

The fibres in all muscles are organized into bundles (as described later), and the fibres in each bundle run parallel with each other. However, the arrangement of the bundles of fibres with respect to the origin and insertion of the muscle is either pennate or non-pennate. In a pennate muscle, the fibres run obliquely with respect to the origin and insertion so that the line of pull of the fibres is oblique to the line of pull of the muscle (Figure 6.9). Pennate muscles have a feather-like appearance and are classified according to the number of groups of fibres into unipennate, bipennate and multipennate. In a unipennate muscle there is one group of fibres that inserts onto the sides of two non-parallel tendons (or

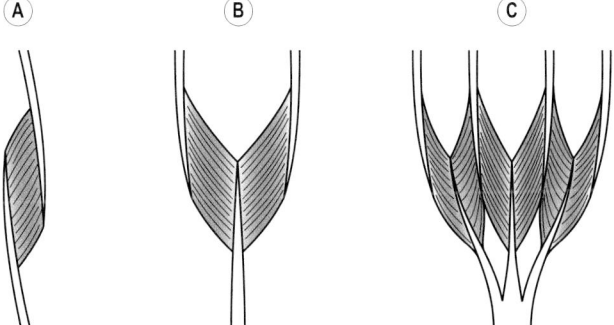

Figure 6.9 Pennate muscles. (A) Unipennate (e.g. finger flexors). (B) Bipennate (e.g. gastrocnemius). (C) Multipennate (e.g. deltoid).

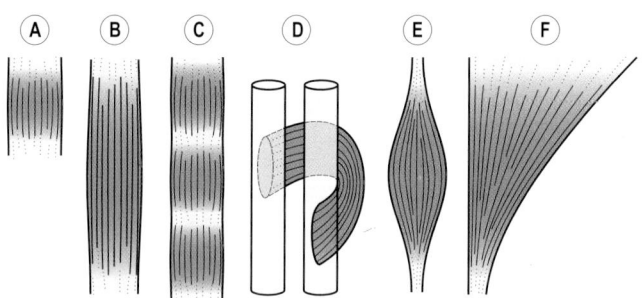

Figure 6.10 Non-pennate muscles. (A) Quadrilateral (e.g. Pronator quadratus). (B) Strap (e.g. sartorius). (C) Strap with tendinous intersections (e.g. rectus abdominis). (D) Spiral (e.g. supinator). (E) Fusiform (e.g. biceps brachii). (F) Fan-shaped (e.g. pectoralis major).

one bony attachment and one non-parallel tendon) (see **Figure 6.9A**). The flexors and extensors of the fingers are unipennate muscles. In a bipennate muscle such as the gastrocnemius there are two groups of fibres that insert onto the opposite sides of a central tendon (see **Figure 6.9B**). A multipennate muscle such as the deltoid is, in effect, two or more bipennate muscles combined into a single muscle (see **Figure 6.9C**).

In a non-pennate muscle, the fibres run in line with the line of pull of the muscle. There are five main types: quadrilateral, strap, spiral, fusiform and fan shaped (**Figure 6.10**). In a spiral muscle, the muscle curves around other muscles or bones. In a fusiform (or spindle-shaped) muscle, the fibres are gathered at each end to attach onto long, relatively narrow tendons. In a fan-shaped muscle, the fibres converge from a broad origin to a relatively small insertion. The effect of pennate and non-pennate arrangements on muscle function is described later in the chapter.

Fusiform muscle–tendon units

Figure 6.11 shows the structure of a fusiform muscle–tendon unit. Each muscle fibre is enveloped in a layer of areolar tissue called the endomysium, which helps to bind the muscle fibres together and provides a supporting framework for blood capillaries and the terminal branches of nerve fibres. The muscle fibres are grouped together by irregular connective tissue (a mixture of collagenous and elastic) in bundles of up to 200 fibres; each bundle is called a fasciculus (or funiculus) and the connective tissue sheath is called the perimysium. The fasciculi are bound together to form the belly of the muscle by a layer of irregular collagenous connective tissue called the epimysium. The muscle fibres gradually taper at each end. The

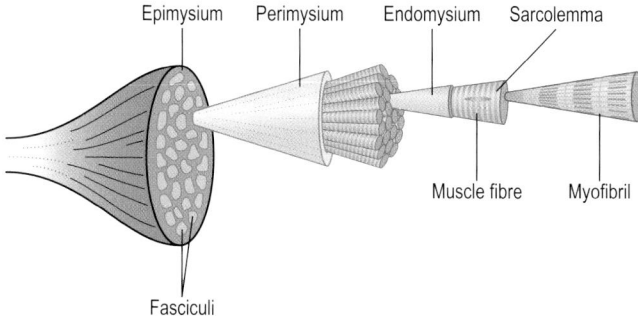

Epimysium Perimysium Endomysium Sarcolemma

Muscle fibre Myofibril

Fasciculi

Figure 6.11 Structure of a fusiform muscle.

tapering of the fibres accompanies a gradual thickening in the epimysium and perimysium layers and a change in the composition of the epimysium and perimysium layers from irregular collagenous and irregular elastic connective tissue to regular collagenous connective tissue as the epimysium and perimysium layers merge to form a tendon or aponeurosis.

Each muscle receives one or more nerves (collections of sensory and motor nerve fibres), which usually enter the muscle together with the main blood vessels (arteries enter, veins leave) at a region of the muscle that does not move a great deal during normal movement; this region is referred to as the neurovascular hilus (Williams et al 1995). The blood vessels and nerves branch through the epimysium and perimysium layers down to the endomysium of the individual muscle fibres.

Key Concepts

A skeletal muscle is made up of skeletal muscle fibres bound together by various layers of connective tissue that support extensive networks of nerves and blood vessels and form tendons and aponeuroses. The muscle fibres are organized into bundles that have a pennate or non-pennate arrangement

Muscle fibres

A muscle fibre consists of hundreds or thousands of myofibrils embedded in sarcoplasm (muscle cytoplasm) and enclosed by a cell membrane called the sarcolemma (**Figure 6.12A and B**). Each muscle fibre has a large number of nuclei that lie just beneath the sarcolemma. Each myofibril is approximately 1 μm wide; the myofibrils are arranged parallel

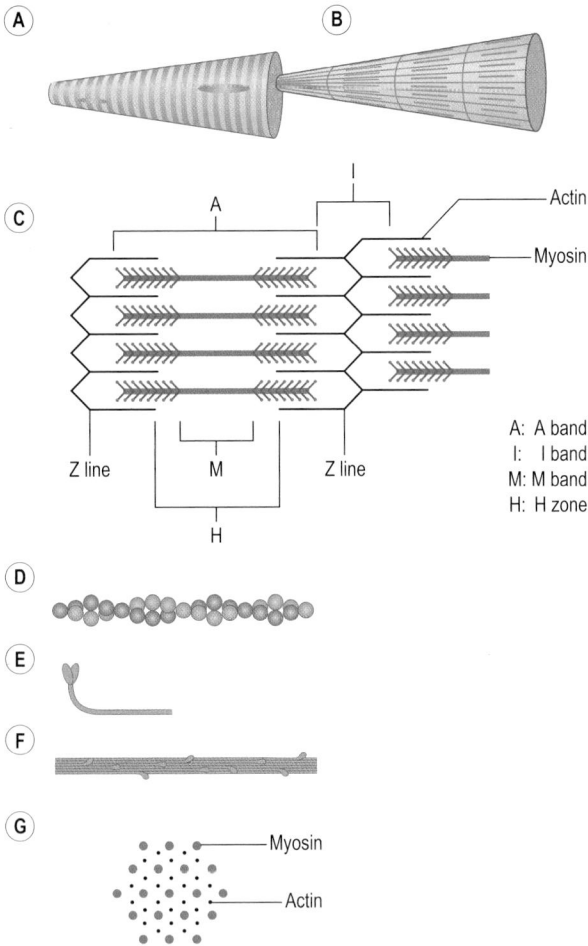

Figure 6.12 Structure of a muscle fibre. (A) Muscle fibre. (B) Myofibril. (C) Arrangement of actin and myosin filaments in a sarcomere. (D) Part of an actin filament. (E) Myosin molecule. (F) Part of a myosin filament. (G) Hexagonal arrangement of actin and myosin filaments in a sarcomere.

to each other and run the whole length of the muscle fibre. Each myofi-bril exhibits a characteristic pattern of alternate light and dark transverse bands due to the way that the components of the myofibril reflect light (under an electron microscope). The light and dark bands are referred to as I (isotropic) and A (anisotropic) bands, respectively. Since the light and dark bands coincide in adjacent myofibrils, the muscle fibre has a striped or striated appearance; thus, skeletal muscle is often referred to as striated muscle.

Each myofibril has two types of filaments arranged in a highly ordered way that gives rise to the light and dark bands. One type of filament is thicker than the other. The thicker filaments occupy the A bands and are composed of the protein myosin; these filaments are usually referred to as myosin filaments or A filaments (see Figure 6.12C). The thinner filaments, actin filaments or I filaments, occupy the I bands and are composed largely of the protein actin (Edman 1992). Each I band is divided by a transverse Z disc. The section of a myofibril between two successive Z discs is called a sarcomere, which is the basic structural unit of a muscle fibre. A myofibril consists of a chain of sarcomeres. The actin filaments project from each side of the Z discs and reach into the A bands of the corresponding sarcomeres where they interdigitate with the myosin filaments (see Figure 6.12C). The region between the ends of the two groups of actin filaments in the middle of a sarcomere is referred to as the H zone.

Each actin filament basically consists of two strands of actin molecules wound together longitudinally in a helical manner (see Figure 6.12D). Each myosin filament is composed of myosin molecules. Each myosin molecule is a club-like structure consisting of two adjacent globular heads attached by a relatively short curved neck to a long shaft (see Figure 6.12E). In a myosin filament, the myosin molecules are packed such that the shafts form the main body of the filament with the heads projecting from the main body at regular intervals (see Figure 6.12F). The two halves of each myosin filament are mirror images of each other, i.e. the myosin molecules in one half of the filament are oriented in the opposite direction to the myosin molecules in the other half. This arrangement, significant in the functional interaction between the actin and myosin filaments, is such that no myosin heads project from the central region of each filament; this central region is sometimes referred to as the inert zone, M band or M line. Each myosin filament is surrounded by six actin filaments in a regular hexagonal arrangement (see Figure 6.12G). In each half of a myosin filament, the myosin molecules are arranged in groups of six; a line joining the heads of the molecules in each group forms a spiral around the myosin filament. The corresponding head in each group faces the same actin filament.

Types of muscular contraction

When a muscle fibre contracts, the heads of the myosin molecules attach to special sites on the actin filaments in the regions where the actin and myosin filaments overlap. The attachments are referred to as cross bridges and each cross bridge generates a certain amount of tension. When a muscle fibre relaxes, the myosin heads detach from the actin and no tension is exerted between them.

When a muscle contracts, the tension exerted by the muscle is directly proportional to the number of cross bridges; the larger the number of cross bridges, the greater the tension. When a muscle contracts it tends to shorten, i.e. it tends to pull its origin and insertion closer together. However, while contracting, the muscle may shorten, lengthen or stay the same length depending on the external load on the muscle (the load tending to lengthen the muscle). If the muscle force is greater than the external load, the muscle will shorten; this type of contraction is called a concentric contraction. If the muscle force is less than the external load, the muscle will lengthen; this type of contraction is called an eccentric contraction. Concentric and eccentric contractions are often referred to as isotonic contractions, i.e. contractions that involve a change in the length of the muscle. If the muscle force is equal to the external load, the length of the muscle will not change; this type of contraction is called an isometric contraction.

Sliding filament theory of muscular contraction

During all types of muscular contraction the actin and myosin filaments stay the same length, but in isotonic contractions the degree of interdigitation between the two sets of filaments changes as the length of the muscle fibres changes; the width of the A bands stays the same, but the width of the I bands and H zones varies. As the muscle fibres shorten, the region of interdigitation increases and the width of the I bands and H zones decreases. As the muscle fibres lengthen, the region of interdigitation decreases and the width of the I bands and H zones increases. These observations led to the formulation of the sliding filament theory of muscular contraction (Huxley and Hanson 1954). The essential features of the theory are as follows:

1. When a muscle contracts, force is generated by the formation of cross bridges.
2. Flexion of the necks of the myosin molecules while the heads are in contact with the actin filaments exerts a pulling force on the actin filaments, which causes the actin filaments to slide relative to the myosin filaments, so that the muscle shortens.
3. After exerting their pulling action, the heads of the myosin molecules detach (or decouple) from the actin filaments and swing back to

reattach (or recouple) onto the actin filaments further along the actin filaments.

4. The coupling and decoupling of different cross bridges occurs at different times so that tension can be maintained while the muscle fibre shortens.

Whereas details of the processes involved have still to be discovered, the sliding filament theory has now gained general acceptance (Edman 1992).

Motor units

The functional unit of skeletal muscle is the motor unit. A motor unit consists of a motor neurone with an Aα axon (sometimes referred to as an alpha motoneurone), together with all the terminal branches of the axon and the muscle fibres that they innervate (Gamble 1988) (Figure 6.13). The number of muscle fibres innervated by a single alpha motoneurone is referred to as the innervation ratio. The innervation ratio of different motor units varies considerably, from approximately 1:4 in the muscles that move the eyes, to approximately 1:2000 in the large back extensor and leg extensor muscles. When an alpha motoneurone transmits an action potential, all of the fibres in the motor unit contract. Consequently, muscles associated with very fine motor control, such as the muscles that move the eyes, have motor units with low innervation ratios so that the amount of force produced can be finely controlled. Muscles associated with forceful movements are made up of motor units with relatively high innervation ratios. The fibres of a single motor unit are usually mixed with the fibres of other motor units, but are grouped within a relatively small area of the muscle.

Key *Concepts*

The innervation ratio is the number of muscle fibres in a motor unit. Muscles associated with fine motor control have motor units with low innervation ratios, whereas muscles associated with forceful movements are made up of motor units with relatively high innervation ratios

A muscle fibre contracts when an action potential is generated in the sarcolemma at the junction with a motor end plate. The action potential is transmitted along and across the muscle fibre (by a system of tubules

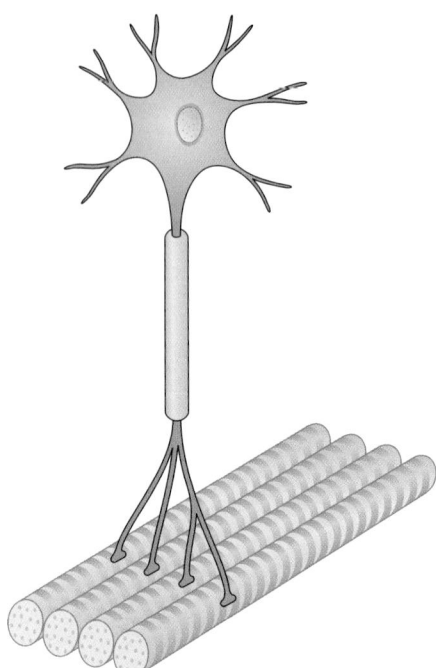

Figure 6.13 Composition of a motor unit: an alpha motoneurone with all the terminal branches of the axon and the muscle fibres that they innervate.

continuous with the sarcolemma) resulting in contraction. A single action potential produces a muscle twitch, i.e. a force that rapidly peaks and then equally rapidly dies away. If a series of action potentials is generated at a high enough frequency, the twitches fuse to produce tetanus, a sustained level of force.

Slow and fast twitch muscle fibres

Whereas the basic structure and function of all muscle fibres is the same, muscle fibres and, consequently, motor units vary in relation to the following:

1. Activation threshold: the level of stimulus required to generate an action potential.
2. Contraction time: the time from force onset to peak force.
3. Resistance to fatigue.

Muscle fibres are classified into slow twitch fibres and fast twitch fibres on the basis of their contraction times. Slow twitch fibres, also referred to as type I and red fibres, have contraction times of 100 to 120 ms. Fast twitch fibres, also referred to as type II fibres and white fibres, have contraction times of 40 to 45 ms (Gamble 1988; Gregor 1993). The metabolism of slow twitch fibres is basically aerobic and, therefore, they are resistant to fatigue. Fast twitch fibres are subdivided on the basis of their metabolic characteristics into type IIa (aerobic and fatigue resistant) and type IIb (anaerobic and fatigue sensitive). The muscle fibres in a particular motor unit have the same functional characteristics and, as such, motor units can be classified into three categories:

- slow contracting, fatigue resistant (S)
- fast contracting, fatigue resistant (FR)
- fast contracting, fatigable (FF)

Individuals differ in the proportions of the different types of muscle fibres in their muscles. The average person has approximately 50% type I, 25% type IIa and 25% type IIb fibres in their calf muscles. In comparison, elite distance runners have a much higher proportion of type I fibres, and elite sprinters have a much higher proportion of type IIb fibres (Gamble 1988).

Whereas the classification of muscle fibres into types I, IIa and IIb is widely used, the categories represent ranges of metabolic and functional characteristics rather than discrete categories (Sargeant 1994). The metabolic and functional characteristics of muscle fibres appear to be influenced considerably by the type of innervation the fibres receive. In experimental animals, it has been shown that altering the type of innervation results, over time, in a change in the metabolic and functional characteristics of muscle fibres (Noth 1992). **Table** 6.2 lists the characteristics of slow and fast twitch motor units.

Kinaesthetic sense and proprioception

The central nervous system constantly receives sensory information from a wide variety of sources concerning the different aspects of physiological functioning. Awareness of body position and body movement is provided by a range of sensory organs, in particular, those concerned with the sensations of effort and heaviness, timing of the movement of individual body parts, the position of the body in space, joint positions and joint movements.

The input from these sources contributes to what is referred to as kinaesthetic sense (or kinaesthesia) (**Figure** 6.14). Some aspects of

Table 6.2 Characteristics of slow and fast twitch motor units

	Slow twitch fatigue resistant (S)	Fast twitch fatigue resistant (FR)	Fast twitch fatiguable (FF)
Activation threshold of the muscle fibres	Low	Moderate	High
Contraction time of the muscle fibres (ms)	100–120	40–45	40–45
Innervation ratio of the motor unit	Low	Moderate	High
Types of muscle fibre	I	IIa	IIb
Type of axon	Aβ	Aα	Aα
Diameter of the axon (μm)	7–14	12–20	12–20
Speed (m/s)	40–80	65–120	65–120
Duration and size of force	Prolonged low force	Prolonged relatively high force	Intermittent high force
Type of activity	Long distance running and swimming	Kayaking and rowing	Sprinting, throwing, jumping and weight lifting

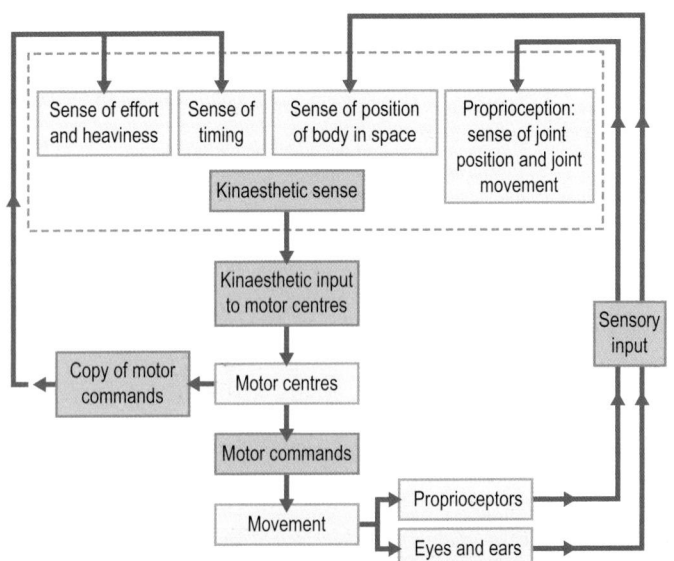

Figure 6.14 Relationship between kinaesthetic sense and proprioception.

kinaesthetic sensitivity, such as a sense of effort and heaviness, and a sense of timing of actions, are generated by sensory centres that monitor the motor commands sent to muscles. Other aspects of kinaesthetic sensitivity are generated largely by input from peripheral receptors that monitor the execution of motor commands, i.e. the actual movements. For example, input from the eyes and ears is responsible for generating a sense of the position of the body in space. The sense of the position of a joint and movement of a joint is generated by a group of receptors located in the skin and musculoskeletal tissues. These receptors are called proprioceptors, and the sensation they provide is called proprioception. Proprioceptors are mechanoreceptors; they are activated by physical distortion.

Proprioceptors

The existence of proprioceptors in the skin, muscles, tendons, joint capsules and ligaments is well established, but the precise role of the different proprioceptors and the interrelationships between them is less clear (Grigg 1994). However, it appears that proprioceptors in joint capsules and ligaments are largely responsible for generating a sense of joint position and joint movement in the end ranges of joint movements, where joint capsules and ligaments may become taut. Between the end ranges it seems unlikely that the proprioceptors in joint capsules and ligaments provide much sensory information since they are unlikely to be under sufficient tension to excite them.

Joint and ligament proprioceptors

There are two main types of proprioceptors in joint capsules, Ruffini end organs (or Ruffini corpuscles) and Pacinian corpuscles. Ruffini end organs, which appear to be mainly responsive to tension and Pacinian corpuscles, which appear to be responsive to compression, are widely distributed between the collagenous fibres of the joint capsule and surrounding fascia.

The proprioceptors in ligaments and skin are similar to those found in joint capsules. However, there are few proprioceptors in ligaments and skin compared to the number in joint capsules. In addition, since tension in ligaments and skin may be caused by movement in a number of different directions, it is unlikely that the proprioceptors can provide information on movement in specific directions. For these reasons, it is thought that the contribution of the proprioceptors in ligaments and skin to proprioception is relatively small (Grigg 1994). However, they may make a significant contribution to joint stabilization in reflexive muscular activity.

Muscle and tendon proprioceptors

Whereas proprioceptors in joint capsules and, to a lesser extent, in ligaments, appear to generate information on joint position and joint movement in the end ranges of joint movements, proprioceptors in muscles, called muscle spindles, and tendons, called Golgi tendon organs, seem to be responsible for generating this type of information for movement between end ranges. Golgi tendon organs and muscle spindles are responsive to tension.

Each muscle spindle consists of a number of tiny muscle fibres enclosed within a spindle-shaped connective tissue capsule (**Figure 6.15**). The muscle spindle fibres are referred to as intrafusal (inside the spindle) to distinguish them from the larger and more numerous extrafusal fibres that make up the vast majority of the muscle. Muscle spindles are embedded between extrafusal fibres. Intrafusal fibres are classified on the arrangement of their nuclei into nuclear bag fibres and nuclear chain fibres. In both types of fibre, the nuclei are located in the central region of the fibre, which is devoid of myofibrils. In nuclear bag fibres, the nuclei cluster in a group, whereas in nuclear chain fibres the nuclei are arranged in a line parallel to the long axis of the fibre. The regions of the fibres on each side of the central region contain a large number of myofibrils and are referred to as the polar regions (Enoka 1994).

Nuclear bag fibres and nuclear chain fibres are usually about 8 mm and 4 mm long, respectively, and in a typical muscle spindle there are two bag fibres and four chain fibres (Roberts 1995). The central regions of both types of fibre are supplied with type Ia and type II sensory nerve endings. The endings of type Ia nerve fibres spiral around the intrafusal fibres and the endings of type II nerve fibres consist of a number of branches with end bulbs; the spirals and end bulbs adhere closely to the sarcolemma of each fibre.

The polar regions of the intrafusal fibres are supplied with motor nerve endings from $A\beta$ and $A\gamma$ motor nerve fibres. Stimulation of the intrafusal fibres via these nerves results in contraction of the polar regions, which stretches the central regions and excites the spiral and end bulb nerve endings. This results in sensory discharge via the type Ia and II sensory nerve fibres. The type Ia and II fibres synapse in the spinal cord directly with the $A\alpha$ motor neurones that supply the extrafusal muscle fibres of the same muscle, resulting in contraction of the extrafusal fibres. The level of contraction of the extrafusal fibres depends on the level of activation, which, in turn, depends on the degree of tension in the intrafusal fibres. Under resting conditions there is always a certain amount of tension in the intrafusal muscle fibres (due to activation by the sensory centers of the brain via the $A\beta$ and $A\gamma$ fibres), which, in turn, results in a

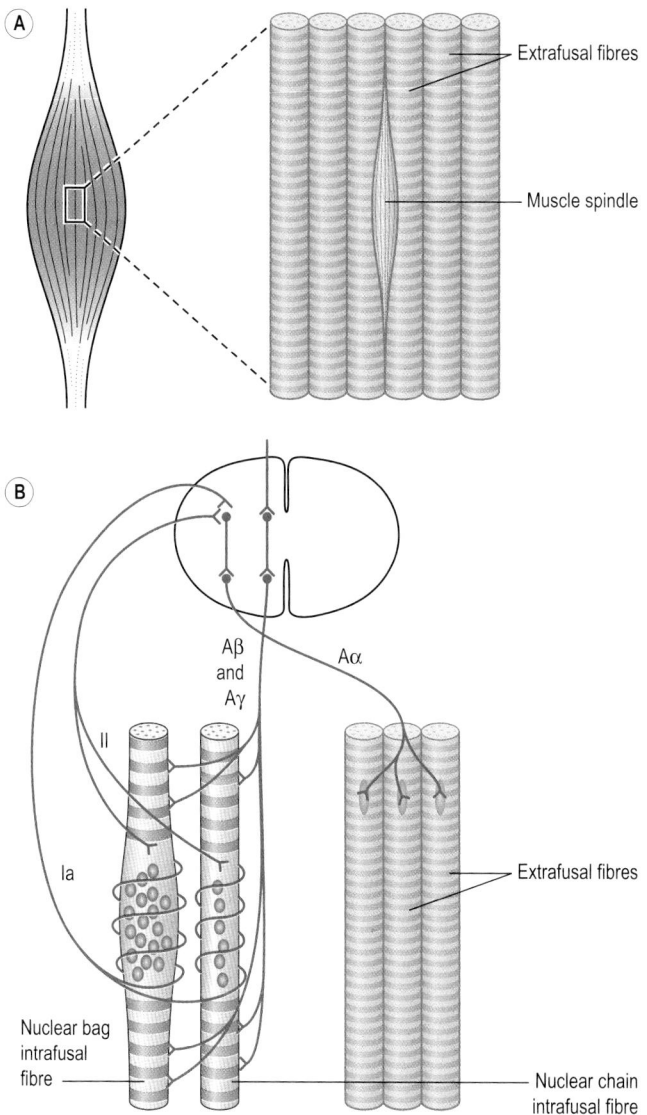

Figure 6.15 Structure of a muscle spindle. (A) Location of a muscle spindle. (B) Innervation of a muscle spindle.

certain amount of tension in the extrafusal muscle fibres. The resting level of tension in the intrafusal and extrafusal muscle fibres is called muscle tone; the level of muscle tone in the extrafusal fibres is determined by the level of muscle tone in the intrafusal fibres.

Sensory output from muscle spindles can be generated by stimulation via the Aβ and Aγ motor nerve fibres (sometimes referred to as the fusimotor nerves) that innervate the intrafusal fibres. However, sensory output from muscle spindles can also be generated by stretching the muscle as a whole, since this will stretch the muscle spindles and excite the spiral and end bulb nerve endings. Low-velocity stretching of active muscles, eccentric muscle contraction, is an essential feature of normal movements; as a muscle group shortens, its antagonist partner experiences eccentric contraction. It is thought that the sensory information provided by muscle spindles as a result of stretching during eccentric contraction provides a sense of joint position and joint movement, especially during mid-range movements (Gandevia et al 1992). The amount of sensory information generated depends on the muscle tone of the intrafusal fibres; increasing the muscle tone of the intrafusal fibres increases the sensitivity of the spindles to stretch, and vice versa.

Whereas low-level stretch of a muscle seems to be important in generating proprioceptive information concerning joint movement and joint position, rapid stretch results in reflex contraction of the muscle (via the spindle afferent to muscle efferent loop) to prevent subluxation of the associated joints. These stretch reflexes are important in protecting joints from injury. For example, recurrent inversion sprains of the ankle are associated with deficient stretch reflex of the everters and dorsiflexors (Garn & Newton 1988). Excitation of muscle spindles appears to be mainly responsible for initiating reflex muscle contractions. However, there is evidence that proprioceptors in joint capsules and ligaments may also contribute to the initiation of such reflex contractions, especially at the ends of joint ranges of movement (Matthews 1988; Hall et al 1994). In these situations, joint capsules and ligaments can be taut and their proprioceptors most active.

Key *Concepts*

Proprioceptors in joint capsules and ligaments (Ruffini end organs and Pacinian corpuscles) appear to generate information on joint position and joint movement in the end ranges of joint movements. Proprioceptors in muscles (muscle spindles) and tendons (Golgi tendon organs) appear to be responsible for generating this type of information for movement in between end ranges of joint movements

Role of proprioceptors

The precise roles of the various types of proprioceptors are not yet clear. However, there is general agreement that proprioceptive information aids coordination and balance and, in particular, maintains joint congruence (Grigg 1994; Wilkerson & Nitz 1994). It appears that injury to muscles, ligaments and capsules may damage proprioceptors resulting in long-term proprioceptive deficits, which, in turn, contribute to the development of degenerative joint diseases such as osteoarthritis (Freeman & Wyke 1967; Garn & Newton 1988; Hall et al 1995). There is evidence that proprioceptive information from muscle–tendon units can be enhanced by specific exercises that emphasize activation of muscle spindles and, thereby, improve muscle tone. Such enhancement may compensate for proprioceptive deficits in other structures such as joint capsules and ligaments (Skinner et al 1986; Steiner et al 1986; Beard et al 1994).

Mechanical characteristics of muscle–tendon units

The amount of force generated by a muscle–tendon unit depends on the length of the unit at the time of stimulation and the speed with which it changes length in the ensuing contraction. In this regard, an isometric contraction is simply one point on the continuum between maximum velocity of shortening (concentric contraction) and maximum velocity of lengthening (eccentric contraction).

Length–tension relationship in a sarcomere

The amount of tension generated in a sarcomere depends on the number of cross bridges attached between the actin and myosin filaments; the greater the number of cross bridges, the greater the tension. The actual number of cross bridges depends on the degree of interdigitation between the actin and myosin filaments. Too much and too little interdigitation decreases the number of myosin heads that are in a position to attach to form cross bridges.

Figure 6.16 shows the isometric length–tension relationship for a sarcomere; the sarcomere was maximally stimulated, but not allowed to shorten, at a number of different sarcomere lengths and the force was recorded at each length (Gordon et al 1966). When the sarcomere is extended to the point where there is no interdigitation between the actin and myosin filaments, no tension is generated since there are no myosin heads in a position to attach to form cross bridges. This situation is represented by the point L5 in Figure 6.16. As the sarcomere shortens and

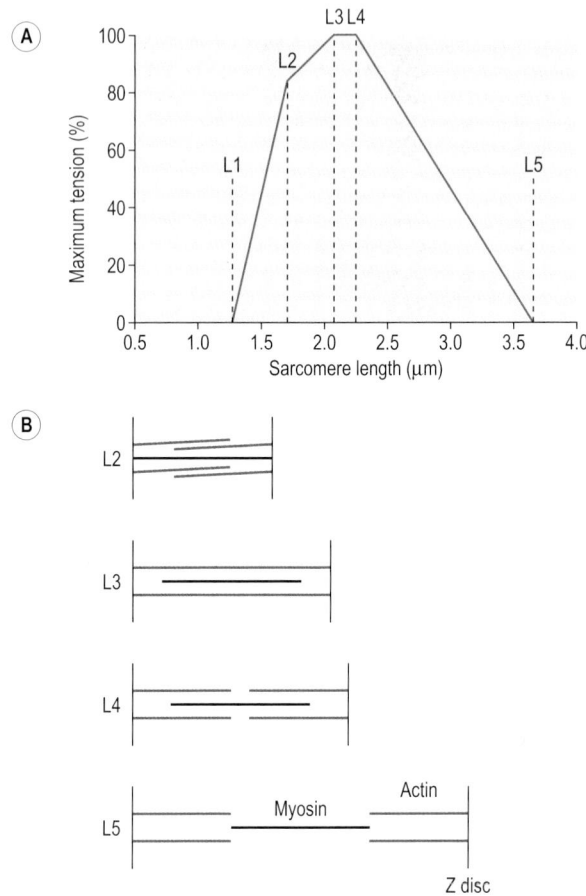

Figure 6.16 Isometric length–tension relationship in a sarcomere.

the degree of interdigitation gradually increases, there is a linear increase in the amount of force generated up to the point L4 where the length–tension relationship levels off. At L4 the maximum number of cross bridges has been formed and, as such, tension is maximum.

The region of the length–tension relationship between L4 and L5 is referred to as the descending limb. At L4 the H zone corresponds to the inert zone of the myosin filaments (see **Figure 6.12C**). As the sarcomere shortens between L4 and L3 the tension stays the same since no more cross bridges can be formed (the inert zones of the myosin filaments do not have myosin heads to attach cross bridges). The region of the length–tension relationship between L3 and L4 is referred to as the plateau region. At L3 the ends of the actin filaments in each half of the sarcomere come together, i.e. the H zone is zero. As the sarcomere shortens between L3 and L2 the actin filaments progressively overlap each other, resulting in a progressive drop in tension since the overlapping actin filaments interfere with cross bridge formation in the region of overlap. The region of the length–tension relationship between L2 and L3 is referred to as the shallow ascending limb. At L2 the ends of the myosin filaments abut the Z discs. As the sarcomere shortens between L2 and L1 there is a rapid and progressive drop in tension due to progressive overlap of the actin filaments and progressive longitudinal compression of the myosin filaments, both of which reduce the number of myosin heads available to form cross bridges. At L1 no cross bridges can be formed and, consequently, no tension is generated. The region of the length–tension relationship between L1 and L2 is referred to as the steep ascending limb.

Length–tension relationship in a muscle–tendon unit

The length–tension relationship of a muscle–tendon unit is different from that of a sarcomere due to the muscle–tendon unit's connective tissue, which exerts passive tension when stretched. Consequently, the tension produced by a muscle–tendon unit will be the sum of the tension produced by the contractile (muscle) component and the passive tension exerted by the connective tissue components. **Figure 6.17** shows the isometric length–tension relationship (or curve) for a muscle–tendon unit. The contributions of the contractile and connective tissue components to total tension at any particular length are shown in the separate curves. Some of the connective tissue components are parallel with the muscle fibres and some are arranged in series; this has given rise to the terms parallel elastic component and series elastic component (Huijing 1992). The parallel elastic component consists of sarcolemma, endomysia, perimysia and epimysium. The series elastic component consists of tendons and aponeuroses, and strands of the protein titin, which connect the ends of the myosin filaments to the Z discs in each sarcomere. In addition to contributing to passive tension when stretched, the titin strands stabilize the hexagonal arrangement of the actin and myosin filaments (Lieber and Bodine-Fowler 1993).

In the absence of stimulation and any external load, the muscle–tendon unit assumes a rest length at which the tension in the unit is zero,

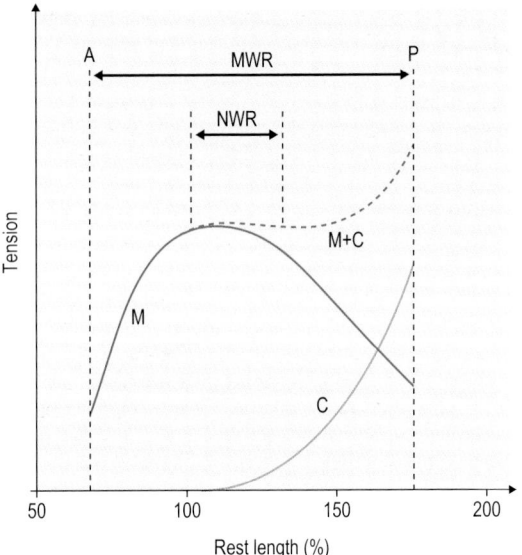

Figure 6.17 Isometric length–tension relationship in a muscle–tendon unit. M, tension exerted by the contractile component; C, tension exerted by the connective tissue component; M + C, total tension; P, passive insufficiency; A, active insufficiency.

with no tension in the contractile component, and no tension (no stretch) in the connective tissue component. Figure 6.17 shows that rest length is associated with that part of the isometric length–tension curve at which tension in the contractile component is maximum.

As in a sarcomere, the shorter the muscle–tendon unit, the lower the tension generated. Tension tends to reduce to zero when the unit shortens to approximately 60% of its rest length (see Figure 6.17). However, this is unlikely to occur in practice due to the arrangement of the muscle–tendon units on the skeleton. When a muscle-tendon unit contracts in a very short-ened state and, as such, generates low force, it is said to be in a state of active insufficiency (Elftman 1966). This is more likely to occur with muscle–tendon units that cross more than one joint. For example, the hamstrings extend the hip and flex the knee (see Figure 2.5). However, the muscle fibres in the hamstrings are not long enough to fully extend the hip and flex the knee simultaneously. If you stand on one leg and attempt to fully extend the hip and fully flex the knee of the other leg at the same time, you will be unable to flex the knee much more than 90°; i.e. the hamstrings will be in a state of active insufficiency. In contrast, if you flex the hip and then flex the knee, you will be able to flex the knee approximately 140°. This is possible because the hamstrings operate closer to rest length and are therefore able to exert more force (Elftman 1966).

Figure 6.17 shows that as a muscle–tendon unit lengthens beyond rest length, the isometric tension generated is fairly constant between 100% and 150% of rest length and then increases to a maximum at approximately 175% of rest length. The change in isometric tension between rest length and maximum tension is associated with a gradual decrease in the amount of tension produced by the contractile component and a gradual increase in the amount of passive tension. Tension in the contractile component would be reduced to zero if the muscle–tendon unit were lengthened to approximately 210% of rest length. However, the parallel elastic connective tissue components ensure that this situation does not arise by limiting the maximum length of the muscle–tendon unit to approximately 175% of rest length. In this situation, the muscle–tendon unit is said to be in a state of passive insufficiency (Elftman 1966).

When a muscle–tendon unit lengthens, all of the sarcomeres do not lengthen to the same extent. Theoretically a situation could occur where some of the sarcomeres in a muscle fibre were fully extended, i.e. no interdigitation between the actin and myosin filaments, while other sarcomeres were not fully extended. If the muscle were stimulated to contract in this state, the fully stretched sarcomeres would not be able to contract, whereas the other sarcomeres would be able to contract. This would result in further stretching and, consequently, damage to the already fully stretched sarcomeres. This theoretical condition has been referred to as muscle instability, and may be prevented, at least in part, by passive insufficiency (Alexander 1989). Consequently, the maximum working length range of a muscle–tendon unit is determined by the lengths at which active insufficiency and passive insufficiency occur, i.e. between approximately 75% and 175% of rest length. However, it is likely that the normal working range is between approximately 100% and 130% of rest length. This range incorporates the region of the length–tension curve where contractile tension is maximum and, as such, allows maximum flexibility in tension generation (see Figure 6.17).

Force-velocity relationship in a muscle–tendon unit

The everyday physical tasks that individuals perform are usually well within the strength capability of the muscle–tendon units used. In such movements, the muscle–tendon units generate just enough tension to overcome the external load acting on them so they can move the external load. The external load may simply be the weight of a limb segment such as the forearm in a movement involving elbow flexion. At other times the external load consists of the weight of the limb segments together with any additional load that is being moved such as something held in the hand.

When the amount of force produced by a muscle (muscle–tendon unit) just matches the external load, the muscle contracts isometrically. The maximum load the muscle can sustain isometrically is called the iso-metric strength of the muscle. When the external load is less than iso-metric strength, the muscle is able to contract concentrically. The speed of shortening in a concentric contraction depends on how much force the muscle needs to produce to move the external load. The greater the external load, the greater the muscle force needs to be, and the greater the muscle force (as a proportion of isometric strength), the slower the speed of shortening. A muscle can shorten at maximum speed when the external load on the muscle is zero. When the external load on a muscle is greater than the isometric strength of the muscle, it is forced to lengthen (contract eccentrically).

In an eccentric contraction a muscle resists the stretching load. In so doing, the attached cross bridges are themselves stretched, adding to the overall tension such that the force produced by the muscle is greater than the isometric strength of the muscle. The force produced by a muscle during eccentric contraction depends on the speed of lengthening, which depends on the size of the external load. The greater the external load (in relation to the isometric strength of the muscle), the greater the speed of lengthening. The greater the speed of lengthening, the greater the effect of the stretch reflex, and, therefore, the greater the force produced by the muscle. When the external force exceeds the maximum strength of the muscle, the muscle and its tendon will be damaged. The relationship between muscle force and speed of shortening or lengthening is referred to as the force–velocity relationship (**Figure 6.18**). **Figure 6.19** shows the effect of the force–velocity relationship on the length–tension relationship of a muscle–tendon unit. The figure shows that at any particular length, the greater the speed of shortening, the lower the tension, and the greater the speed of lengthening, the higher the tension.

Key Concepts

The amount of force generated by a muscle–tendon unit depends on the length of the muscle–tendon unit at the time of stimulation (length–tension relationship) and the speed with which it changes length in the ensuing contraction (force–velocity relationship)

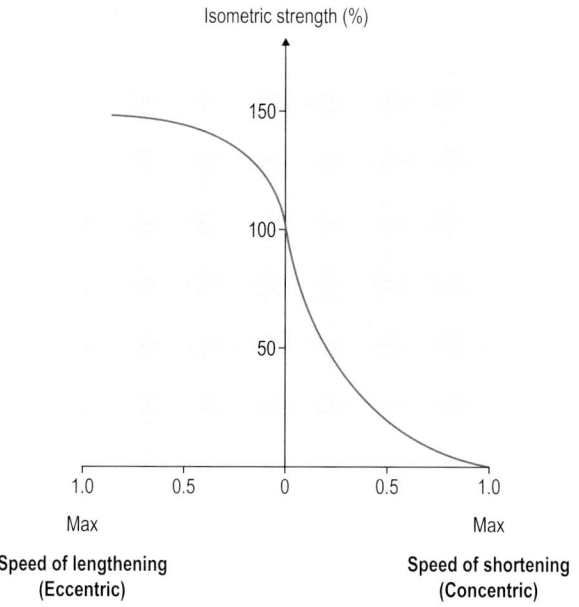

Isometric strength (%)

Figure 6.18 The force–velocity relationship in skeletal muscle.

Muscle architecture and function

All muscles are made up of muscle fibres. However, the length and the orientation of the fibres (pennate or non-pennate) have a considerable effect on the function of the muscles. The fundamental relationships between muscle architecture and muscle function are (1) excursion (the distance that the muscle can shorten) and velocity of shortening are proportional to fibre length, and (2) force is proportional to the total physiological cross-sectional area of the muscle fibres (Lieber & Bodine-Fowler 1993).

All muscle fibres are comprised of similar sarcomeres, and the number of sarcomeres determines the length of a muscle fibre. Each sarcomere in a muscle fibre is capable of shortening to the same extent as all the other sarcomeres in the muscle fibre. Consequently, the excursion of the muscle fibre is equal to the sum of the excursions of all the individual sarcomeres; the greater the number of sarcomeres, the longer the muscle fibre, the greater the excursion. Excursion and velocity of shortening are

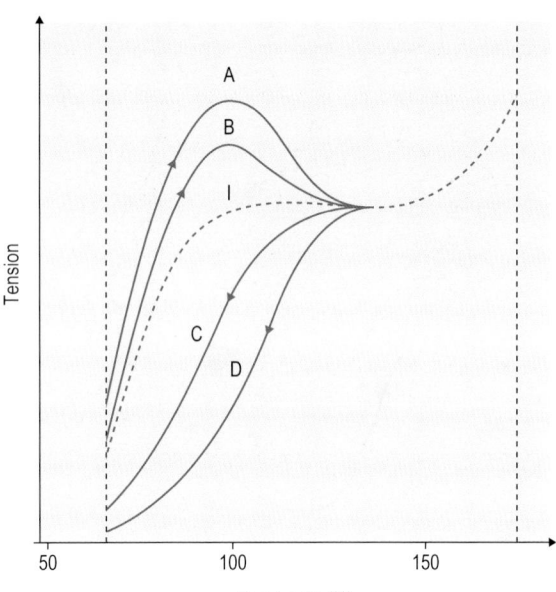

Figure 6.19 The effect of speed of shortening and speed of lengthening on the length–tension relationship in skeletal muscle. A and B show eccentric contractions: the speed of lengthening in **A** > **B**; C and D show concentric contractions: the speed of shortening in C < D; I shows the isometric length–tension curve.

directly related since velocity of shortening is the rate of change of excursion, i.e. the rate of change in length of the muscle. The longer the muscle fibre (in terms of number of sarcomeres), the greater its excursion and velocity of shortening.

Theoretically, the ideal muscle (in terms of force and excursion capabilities) has a large cross-sectional area and very long fibres. However, such a muscle would be bulky and create considerable packing problems due to its girth and areas of attachment to the skeletal system. Since there are no muscles with both of these characteristics, it is reasonable to assume that the architecture of the muscular system has evolved to provide the best compromise between structure and function. The muscles of the body represent a broad range of combinations of force and excursion capability (Lieber 1992), and it is, perhaps, not surprising that most movements of the body involve simultaneous activity in a number of muscles with each muscle performing a particular role.

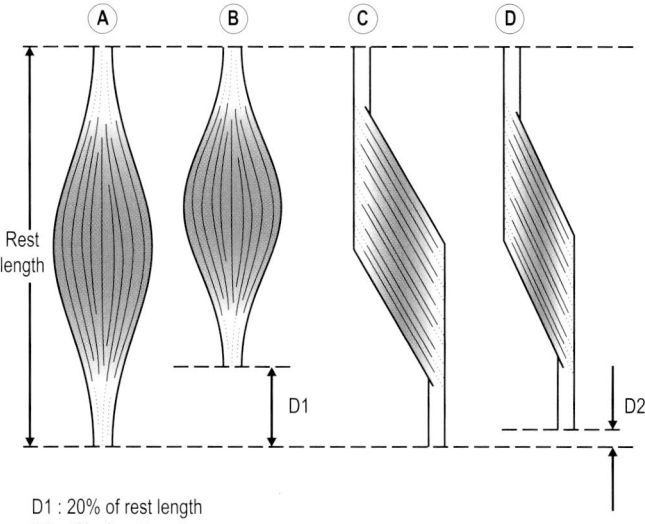

D1 : 20% of rest length
D2 : 5% of rest length

Figure 6.20 Effects of muscle structure on force and excursion. (A) Non-pennate muscle at rest length. (B) Non-pennate muscle in (A) with fibres shortened to 70% of rest length. (C) Pennate muscle at rest length. (D) Pennate muscle in (C) with fibres shortened to 70% of rest length.

Figure 6.20A and C shows two muscle–tendon units with the same rest length and the same muscle mass (same volume of myofibrils). One muscle is a non-pennate parallel-fibred muscle (see Figure 6.20A) and the other is a unipennate muscle (see Figure 6.20C). The physiological cross-sectional area of each muscle is the cross-sectional area of all of the muscle fibres perpendicular to their line of pull. Assuming that the length of the muscle fibres in the pennate muscle is only half that of the fibres in the non-pennate muscle, it follows that the physiological cross-sectional area of the pennate muscle is double that of the non-pennate muscle (since the muscle mass is the same in both muscles). Consequently, the pennate muscle can exert double the force of the non-pennate muscle. However, as the fibres in the pennate muscle are oblique to the line of pull of the muscle–tendon unit, not all of the force is in the line of pull of the muscle–tendon unit. In a pennate muscle, the angle of the fibres with respect to the line of pull of the muscle–tendon unit is usually 30° or less (Alexander 1968), such that 90% or more of the force exerted by the muscle is directed in the line of pull of the muscle. Consequently, pennate muscles are normally capable of exerting far more force in the line of pull of the muscle–tendon unit than non-pennate muscles of the same muscle mass.

In contrast to their capacity to generate force, the excursion range of non-pennate muscle–tendon units is usually much greater than that of pennate muscle–tendon units of similar muscle mass. The increased excursion range of non-pennate muscles is due to increased fibre length in non-pennate muscles and the obliquity of the fibres in pennate muscles, which reduces excursion in the line of pull of the muscle. In **Figure 6.20B and D**, the muscles are shown with their fibres shortened to 70% of their rest length. The corresponding shortening of the muscle–tendon units is 20% and 5% in the non-pennate and pennate muscles, respectively.

Key *Concepts*

The fundamental relationships between muscle architecture and muscle function are that excursion (the distance that the muscle can shorten) and velocity of shortening are proportional to fibre length, and force is proportional to total cross-sectional area of the muscle fibres

Biarticular muscles

Many of the skeletal muscles, especially those in the upper and lower limbs, span more than one joint. These muscles are usually referred to as biarticular muscles; muscles that span more than two joints function in the same way as muscles that span two joints (Lieber 1992). Biarticular muscles are too short to fully flex or fully extend simultaneously all of the joints that they span. For example, the hamstrings are two-joint muscles that can extend the hip and flex the knee, but (as described in a previous section) the hamstrings cannot perform these actions maximally at the same time. Indeed, hip extension is normally associated with knee extension, and hip flexion is normally associated with knee flexion, as in walking and running. In this way, the length of the muscles stay within the normal working range of approximately 100% to 130% of rest length. The functional advantages of biarticular muscles are that tension is produced in one muscle rather than two (or more), which conserves energy, and working within the 100% to 130% range allows maximal flexibility in tension generation (Lieber 1992; Van Ingen Schenau, Pratt & Macpherson 1994).

Review questions

1. Differentiate between the cerebrospinal nervous system and the central nervous system.
2. Differentiate between resting potential and action potential.
3. Describe the general organization of nerve tissue in the spinal cord and spinal nerves.
4. Describe the sequence of events within the nervous system in a typical voluntary movement.
5. Describe the various categories of pennate and non-pennate muscles.
6. Describe the sliding filament theory of muscular contraction.
7. Differentiate between kinaesthetic sense and proprioception.
8. Differentiate between active and passive insufficiency in skeletal muscle.
9. Describe the fundamental relationships between muscle architecture and muscle function.

References

Alexander RMcN (1968) Animal mechanics. Sidgwick and Jackson, London.

Alexander RMcN (1989) Muscles for the job. New Scientist 122(1660):50–53.

Beard DJ, Dodd CAF, Trundle HR, Simpson AHRW (1994) Proprioception enhancement for anterior cruciate ligament deficiency. Journal of Bone and Joint Surgery 76B:654–59.

Edman KAP (1992) Contractile performance of skeletal muscle fibres. In: Komi PV (ed) Strength and power in sport. Blackwell Scientific, Oxford, pp 96–114.

Elftman H (1966) Biomechanics of muscle. Journal of Bone and Joint Surgery 48A:363–77.

Enoka RM (1994) Neuromechanical basis of kinesiology. Human Kinetics, Champaign, IL.

Freeman M, Wyke B (1967) Articular reflexes at the ankle joint: an electromyographic study of normal and abnormal reflexes of ankle joint mechanoreceptors upon reflex activity in the leg muscles. British Journal of Surgery 54:990–1001.

Gamble JG (1988) The musculoskeletal system: physiological basics. Raven Press, New York.

Gandevia SC, McClosky DI, Burke D (1992) Kinesthetic signals and muscle contraction. Trends in Neuroscience 15:62–65.

Garn S, Newton R (1988) Kinesthetic awareness in subjects with multiple ankle sprains. Physical Therapy 68:1667–71.

Gordon AM, Huxley AF, Julian FJ (1966) The variation in isometric tension with sarcomere length in vertebrate muscle fibres. Journal of Physiology (London) 184:170–92.

Gregor RJ (1993) Skeletal muscle mechanics and movement. In: Grabiner MD (ed) Current issues in biomechanics. Human Kinetics, Champaign, IL, pp 171–211.

Grigg P (1994) Peripheral neural mechanisms in proprioception. Journal of Sport Rehabilitation 3:2–17.

Hall MG, Ferrell WR, Baxendale RH, Hamblen DL (1994) Knee joint proprioception: threshold detection levels in healthy young subjects. Neuro-Orthopedics 15:81–90.

Hall MG, Ferrell WR, Sturrock DL, et al (1995) The effect of the hypermobility syndrome on knee joint proprioception. British Journal of Rheumatology 34:121–25.

Huijing PA (1992) Mechanical muscle models. In: Komi PV (ed) Strength and power in sport. Blackwell Scientific, Oxford, pp 130–50.

Huxley HE, Hanson J (1954) Changes in the cross striations of muscle during contraction and stretch and their structural interpretation. Nature 173:973–77.

Kleinrensink GJ, Stoeckart R, Meulstee J, et al (1994) . Medicine and Science in Sports and Exercise 26:877–83.

Lieber RL (1992) Skeletal muscle structure and function. Williams & Wilkins, Baltimore.

Lieber RL, Bodine-Fowler SC (1993) Skeletal muscle mechanics: implications for rehabilitation. Physical Therapy 73:844–56.

Matthews PB (1988) Proprioceptors and their contribution to somatosensory mapping: complex messages require complex processing. Canadian Journal of Physiology and Pharmacology 66:430–38.

Noth J (1992) Motor units. In: Komi PV (ed) Strength and power in sport. Blackwell Scientific, Oxford, pp 21–28.

Roberts TDM (1995) Understanding balance: the mechanics of posture and locomotion. Chapman and Hall, London.

Sargeant AJ (1994) Human power output and muscle fatigue. International Journal of Sports Medicine 15:116–21.

Seddon JH (1972) Surgical disorders of peripheral nerves. Livingstone, Edinburgh.

Skinner HB, Wyatt MP, Stone ML, et al (1986) Exercise-related knee joint laxity. American Journal of Sports Medicine 14:30–34.

Steiner ME, Grana WA, Chillag K, Schelberg-Karnes E (1986) The effect of exercise on anterior-posterior knee laxity. American Journal of Sports Medicine 14:24–29.

Van Ingen Schenau GJ, Pratt CA, Macpherson JM (1994) Differential use and control of mono- and biarticular muscles. Human Movement Science 13:495–517.

Wilkerson GB, Nitz AJ (1994) Dynamic ankle stability: mechanical and neuromuscular relationships. Journal of Sport Rehabilitation 3:43–57.

Williams PL, Bannister LH, Berry MM, et al, eds (1995) Gray's anatomy. Longman, Edinburgh.

Structure and function of the foot

In weightbearing activities, such as standing, walking, running, jumping and landing, the function of the musculoskeletal system is to generate and transmit internal forces to counteract the effects of gravity and create the ground reaction forces necessary to maintain upright posture, transport the body and manipulate objects, often simultaneously, as in running while holding a ball. The ground reaction forces act on the feet and, in general, the greater the speed of movement, the greater the magnitude of the ground reaction forces (Nigg et al 1981; Voloshin 2000). The greater the ground reaction forces, the greater the associated muscle–tendon forces and joint reaction forces. For example, when running at moderate speed (4–5 m/s), peak ground reaction forces will be in the region of 2.5–3.0 BW (body weight). However, the associated Achilles tendon force and ankle joint reaction force will be in the region of 6–8 BW and 10–14 BW, respectively (Scott & Winter 1988). As shown in Chapter 1 (Elementary biomechanics), a relatively small change in the centre of pressure is likely to significantly alter the moments of the ground reaction force about all of the joints of the foot and, consequently, significantly alter the loads on the muscles that counteract the ground reaction force moments. Normally, the path of the centre of pressure along the plantar aspect of the foot does not result in excessive loading on any of the muscles and joints of the foot. However, abnormal passage of the centre of pressure is likely to overload some muscles and joints in the foot which, in turn, as a result of compensatory movements, may result in other parts of the musculoskeletal system, such as the knee, hip and low back, being overloaded.

In dynamic situations, the foot is required to act as both a shock absorber, to cushion the impact of contact of the foot with the ground, and as a propulsive mechanism to propel the body in the desired direction. The foot often performs these functions on a variety of support surfaces. Whereas floor surfaces tend to be firm and level, there are many other situations, such as in cross-country running, where the surface of the ground is neither firm nor level, but continually changes in terms of slope, evenness and hardness. The ability of the foot to function effectively in relation to such diverse environmental constraints is due to its structure, in particular, its flexible arched shape and complex movement capability. The purpose of this chapter is to describe the structure of the foot and the function of the foot during weightbearing.

Rearfoot complex

Many of the 26 bones in each foot articulate with two or more other bones such that there are approximately 40 joints in each foot. Consequently, most movements of the foot involve a large number of joints and the movement of individual joints in each movement is difficult to describe. However, as in most movements of the body, there tends to be high degree of functional interdependence between the joints of the foot, especially between the ankle, subtalar and midtarsal joints, such that movement of one joint tends to bring about fairly predictable movement in adjacent joints (Singh et al 1992; Kitaoka et al 1997a; Nester 1997). A group of joints with a relatively high degree of functional interdependence is called a joint complex (Peat 1986). The term 'rearfoot complex' is frequently used to describe the functional interdependence between the ankle, subtalar and midtarsal joints (Downing et al 1978; Bowden & Bowker 1995; Nester 1997).

Ankle joint

The ankle joint, between the tibia, fibula and talus, is a hinge joint which facilitates rotation about an axis of rotation that runs approximately 20° anterior–superior with respect to the horizontal in the sagittal plane and 20° anterior–medial with respect to the coronal plane in the horizontal plane with respect to the centre of the joint (**Figure 7.1**) (Singh et al 1992).

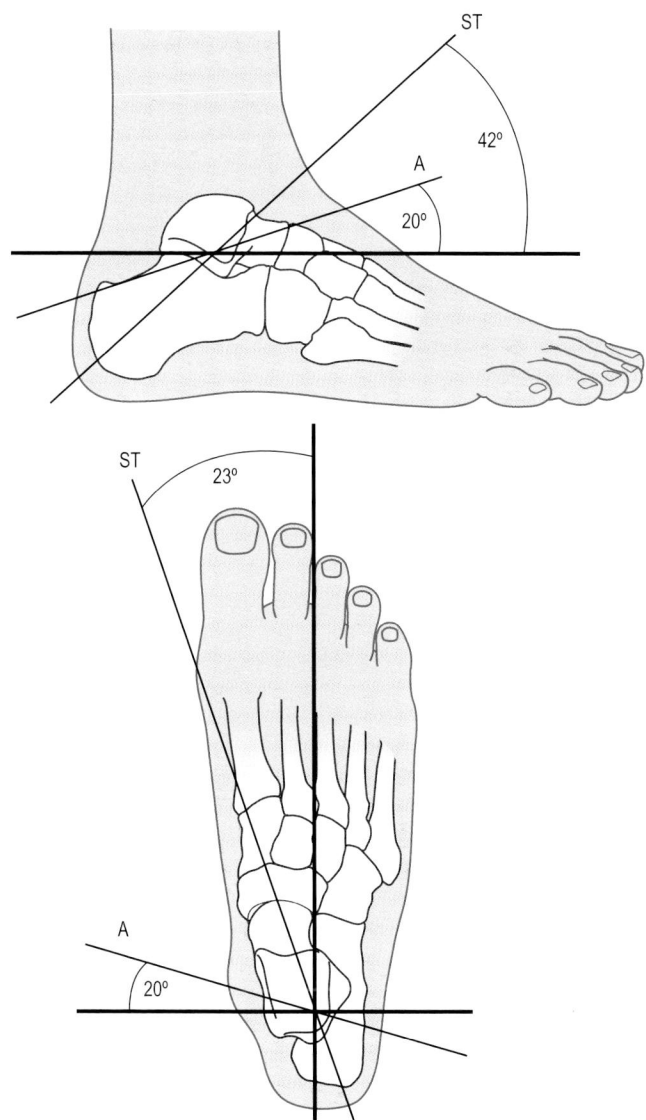

Figure 7.1 Orientation of the axes of rotation of the ankle (A) and subtalar (ST) joints.

Consequently, the movement of the ankle joint is triplanar, i.e. movement occurs simultaneously in the sagittal, coronal and transverse planes about mediolateral, anteroposterior and vertical axes, respectively, with movement predominantly in the sagittal plane. Sagittal plane motion of the foot about the ankle joint is usually referred to as plantar flexion and dorsiflexion (Figure 7.2A). In dorsiflexion, sometimes referred to as true flexion of the ankle, the dorsal (superior) surface of the foot is drawn closer to the shin. In plantar flexion, sometimes referred to as extension of the ankle, the dorsal surface of the foot is moved away from the shin (pointing the toes). Transverse plane motion of the foot about the ankle is usually referred to as abduction and adduction (see Figure 7.2B) and coronal plane motion of the foot about the ankle is usually referred to as eversion and inversion (see Figure 7.2C).

Ankle joint ligaments

The articular surfaces of the distal tibia (trochlea surface and medial malleolus) and distal fibula (lateral malleolus) form a mortise on the talus, which restricts side-to-side movement of the talus in the ankle joint. The mortise is maintained by the integrity of the distal tibiofibular joint, which is a syndesmosis formed by the downward extension and thickening of the interosseous membrane between the shafts of the tibia and fibula (Figure 7.3). The syndesmosis is supported by strong anterior and posterior tibiofibular ligaments. The anterior tibiofibular ligament runs downward and laterally from the anterior lateral aspect of the tibia just above the ankle joint to the anterior aspect of the lateral malleolus. The posterior tibiofibular ligament runs downward and laterally from the posterior lateral aspect of the tibia just above the ankle joint to the posterior aspect of the lateral malleolus (see Figure 7.3). The inferior aspect of the anterior tibiofibular ligament overlaps the anterior lateral aspect of the ankle joint, and the inferior aspect of the posterior tibiofibular ligament overlaps the posterior lateral aspect of the ankle joint.

When viewed from above, the trochlea surface of the talus is wedge shaped with the broad end in front (Figure 7.4). Consequently, the intermalleolar space slightly increases during ankle dorsiflexion, which tightens the mortise (due to strain on the syndesmosis and the anterior and posterior tibiofibular ligaments) and increases the side-side stability of the ankle joint. However, during ankle plantar flexion, the intermalleolar space slightly decreases, which loosens the mortise and tends to reduce the side-side stability of the ankle joint.

The medial aspect of the ankle joint is supported by the deltoid ligament (also referred to as the medial collateral ligament; Figure 7.5). The deltoid ligament is a strong ligament that fans out from the anterior, medial and posterior aspects of the medial malleolus to attach onto a more or

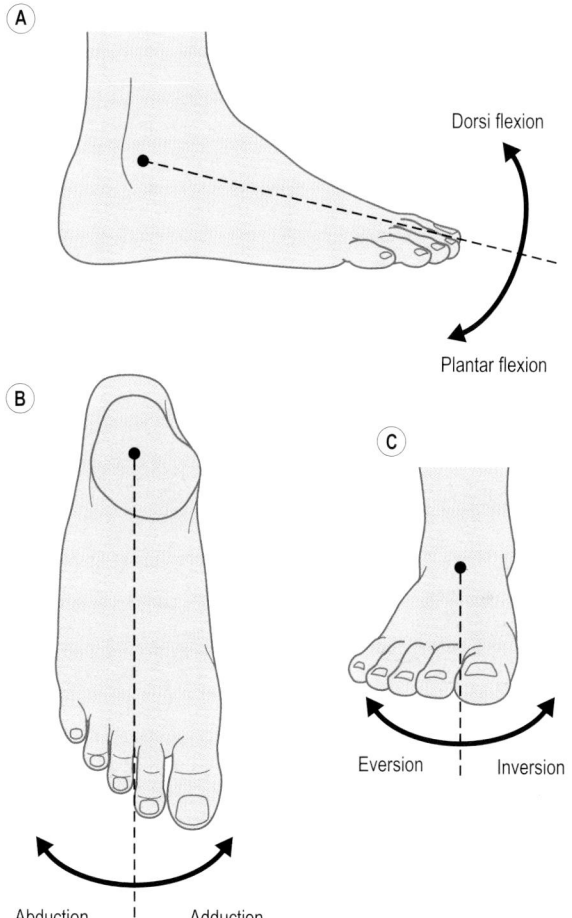

Figure 7.2 Components of triplanar movement of the foot about the axis of the rearfoot complex. (A) Lateral aspect of the right foot showing the mediolateral component *M* of the axis of the rearfoot complex. (B) Superior aspect of the right foot showing the vertical component *V* of the axis of the rearfoot complex. (C) Anterior aspect of the right foot showing the anteroposterior component *A* of the axis of the rearfoot complex.

less continuous arc formed by the tuberosity of the navicular, the plantar calcaneonavicular ligament (which spans the gap between the tuberosity of the navicular and the sustentaculum tali of the calcaneus and supports the head of the talus), the sustentaculum tali and the medial tubercle of

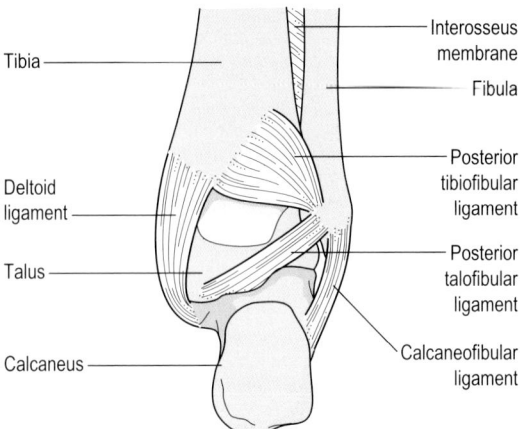

Figure 7.3 Posterior ligaments of the right ankle.

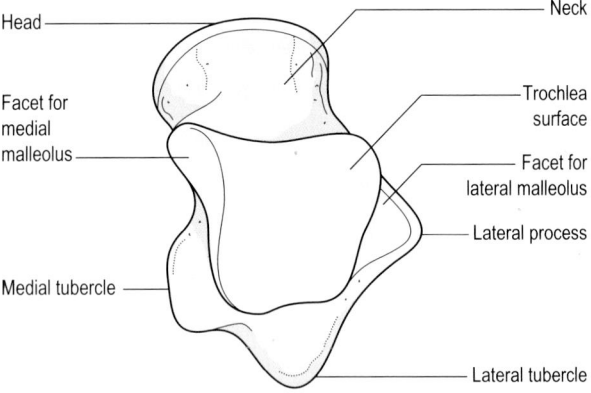

Figure 7.4 Superior aspect of the right talus.

the talus. The deltoid ligament as a whole strongly resists eversion of the ankle and subtalar joints. The fibres of the anterior band become taut at the limit of plantar flexion and the fibres of the posterior band become taut at the limit of dorsiflexion.

The lateral aspect of the ankle joint is supported by the anterior talofibular ligament, the lateral talocalcanean ligament, the calcaneofibular ligament and the posterior talofibular ligament (**Figure 7.6**). The anterior

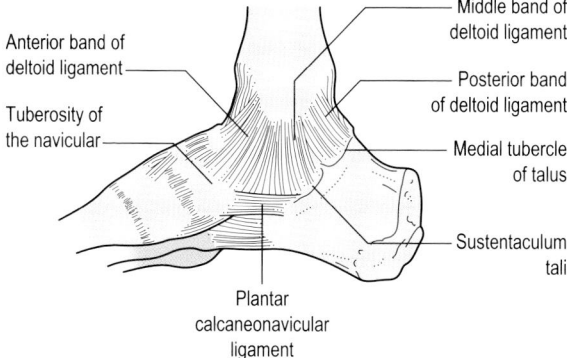

Figure 7.5 Medial ligaments of the right ankle.

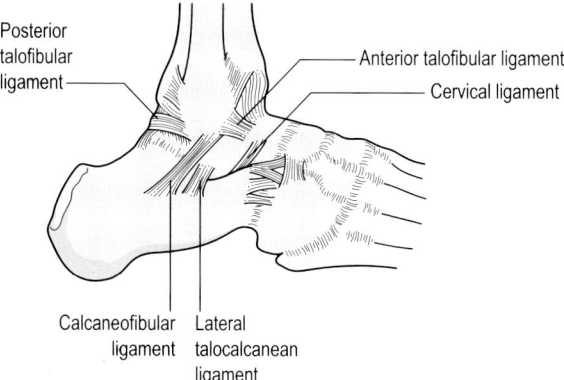

Figure 7.6 Lateral ligaments of the right ankle.

talofibular ligament runs medially and slightly forward and downward from the anterior inferior aspect of the lateral malleolus to the anterior lateral aspect of the talus immediately in front of the lateral articular surface of the talus. The lateral talocalcanean ligament runs downward and slightly backward from the lateral process of the talus (see **Figure** 7.4) to the adjacent lateral aspect of the calcaneus. The calcaneofibular ligament runs medially downward and backward from the inferior aspect of the lateral malleolus to the lateral aspect of the calcaneus. The posterior talofibular ligament runs medially and slightly backward from the posterior inferior aspect of the lateral malleolus to the lateral tubercle of the talus (see **Figure 7.3**). The posterior talofibular ligament becomes taut at the limit of dorsiflexion, and the anterior talofibular ligament becomes taut at the limit of plantar flexion.

Ligament injuries of the ankle are common, especially in sports (Boruta et al 1990). The most common ankle injury is an inversion sprain, which results in damage to one or more of the lateral ligaments and, in some cases, fracture of the medial malleolus or the lateral malleolus (Lassiter et al 1989). Chronic, i.e. persistent lateral ankle instability occurs in 10% to 20% of individuals following a severe inversion sprain (Perlman et al 1987; Peters, Trevino, and Renström 1991). Most inversion sprains involve plantar flexion as well as inversion. As described previously, the ankle joint is least stable in plantar flexion due to loosening of the tibiofibular mortise.

Subtalar joint

The subtalar joint, between the talus and calcaneus, is part synovial and part syndesmosis. The anterior synovial part of the joint is separated from the posterior synovial part of the joint by a funnel-shaped channel called the sinus tarsi. The sinus tarsi runs more or less horizontally in an oblique posterior–medial to anterior–lateral direction (**Figure 7.7**) with the funnel opening out laterally. The posterior talar articular surface of the calcaneus is cylindrical convex and articulates with the reciprocally shaped cylindrical concave posterior calcanean articular surface of the talus. The anterior talar articular surface of the calcaneus (located on the superior aspect of the sustentaculum tali) is cylindrical concave and articulates with the reciprocally shaped cylindrical convex anterior calcanean articular surface of the talus. Whereas **Figure 7.7** shows only one articular surface in the anterior synovial part of the subtalar joint, there are frequently two adjacent articular surfaces. Four distinct variations in the number (one or two), shape and orientation of the anterior synovial articular surfaces have been identified (Valmassy 1996).

The syndesmosis part of the subtalar joint is a broad interosseous talocalcanean ligament consisting of medial and lateral bands that run downward from the sulcus tali (superior part of the sinus tarsi) to the sulcus calcanei (inferior part of the sinus tarsi). In front of the anterior end of the sinus tarsi is another broad ligament called the cervical ligament. The cervical ligament runs obliquely upward and medially from the anterior superior aspect of the calcaneus to the lateral aspect of the neck of the talus (see **Figure 7.6**).

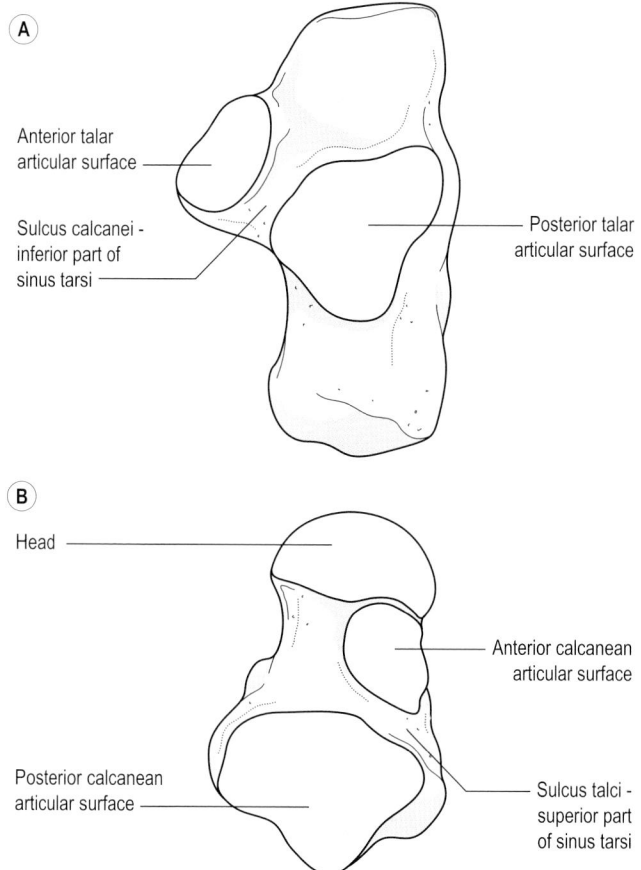

Figure 7.7 Articular surfaces of the right subtalar joint. (A) Superior (dorsal) aspect of the right calcaneus. (B) Inferior (plantar) aspect of the right talus.

Like the ankle joint, the movement of the subtalar joint is triplanar. Inman (1976) showed that the orientation of the axis of the joint varies considerably among individuals with a mean orientation of approximately 42° anterior–superior with respect to the horizontal in the sagittal plane and 23° anterior–medial with respect to the median plane in the horizontal plane with respect to the centre of the joint (see Figure 7.1).

Pronation and supination of the rearfoot complex

In contrast to the ankle and subtalar joints, there would appear to be little empirical information on the movement of the midtarsal joint, which is comprised of a biplanar/biaxial saddle joint (calcaneocuboid) and a triplanar/triaxial ball-and-socket joint (talonavicular). However, it is clear that the rearfoot complex facilitates triplanar movements of the foot. These movements are usually referred to as pronation and supination (**Figure 7.8**) (Kitaoka et al 1997b; Nester 1997).

Pronation involves simultaneous abduction, dorsiflexion and eversion (see **Figure 7.8A and B**). Similarly, supination involves simultaneous adduction, plantar flexion and inversion (see **Figure 7.8B and C**). The orientation of the rearfoot complex axis varies considerably with a mean orientation of approximately 51° anterior–superior with respect to the horizontal in the

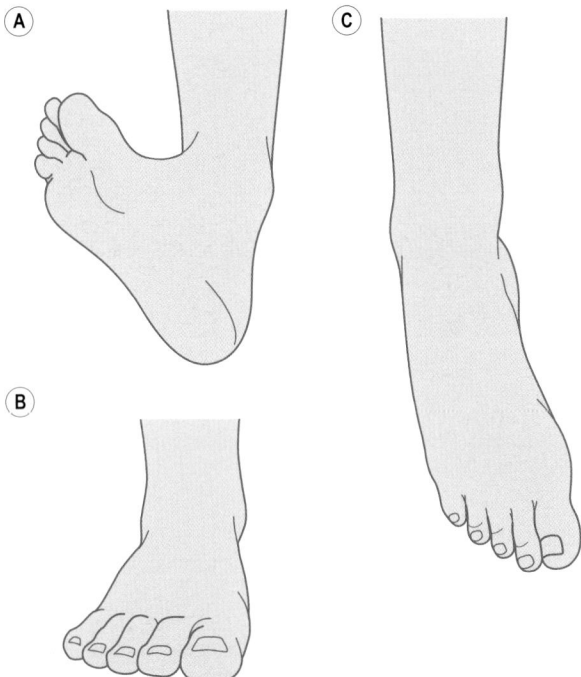

Figure 7.8 Supination and pronation: (A) full pronation; (B) neutral position; (C) full supination.

sagittal plane and 18° anterior–medial with respect to the median plane in the horizontal plane with respect to the centre of the rearfoot complex (Downing et al 1978).

Using 13 cadaver specimens (mean age 65 years, range 20–89 years) and a magnetic tracking measurement system, Kitaoka et al (1997b) investigated the contribution of the ankle, subtalar, talonavicular and the first metatarsal-navicular joints to pronation, supination, dorsiflexion and plantar flexion. The results are shown in Table 7.1. As expected, the ankle is the major contributor (47.2%) to the plantar flexion–dorsiflexion range of motion,

Table 7.1 Contribution of ankle, subtalar, talonavicular and first metatarsal-navicular joints to pronation, supination, dorsiflexion and plantar flexion

	Pronation		Supination		Pronation–supination	
	Degrees	%	Degrees	%	Degrees	%
met-nav	13.6±3.9	43.3	3.3±1.5	4.4	16.9	15.9
tal-nav	7.6±3.3	24.3	39.3±11.8	52.5	46.9	44.2
subtalar	2.5±1.7	8.0	23.3±7.3	31.1	25.8	24.3
ankle	7.6±4.7	24.3	8.9±4.6	11.9	16.5	15.5
ROM	31.3	100	74.8	100	106.1	100
	Dorsiflexion		Plantar flexion		Dorsiflexion–plantar flexion	
	Degrees	%	Degrees	%	Degrees	%
met-nav	1.7±1.1	6.9	11.8±5.5	19.1	13.5	15.6
tal-nav	3.5±1.8	14.3	12.7±8.4	20.6	16.2	18.8
subtalar	2.7±1.7	11.0	6.5±4.2	10.5	9.2	10.7
ankle	16.6±4.8	67.8	30.6±7.9	49.7	47.2	54.8
ROM	24.5	100	61.6	100	86.1	100

ROM, range of motion.
Pronation: from neutral to full pronation; Supination: from neutral to full supination; Dorsiflexion: from neutral to full dorsiflexion; Plantar flexion: from neutral to full plantar flexion; Pronation–supination: range from full pronation to full supination; Dorsiflexion–plantar flexion: range from full dorsiflexion to full plantar flexion.
Data from Kitaoka HB, Luo ZP, An K-N (1997b) Three-dimensional analysis of normal ankle and foot mobility. American Journal of Sports Medicine 25:238–42.

but there are significant contributions from the other joints. The subtalar joint is often regarded as the major contributor to the pronation–supination range of motion, but the results of the study indicate that the contribution of the subtalar joint (24.3%) is less than that of the talonavicular joint (44.2%).

Pronation and supination during weightbearing

The movements of supination and pronation as described above refer to movements of the rearfoot complex when the foot is non-weightbearing. When the foot is weightbearing, these movements are constrained depending on the magnitude and distribution of the ground reaction force acting on the plantar (inferior) surface of the foot. Under weightbearing conditions the most noticeable movements of the foot occur about an anteroposterior axis through the foot (similar to inversion and eversion). For this reason, in describing the movement of the foot under weightbearing conditions the terms supination and inversion are sometimes used synonymously, as are the terms pronation and eversion. However, the actual movements of the foot under weightbearing conditions are modifications of supination and pronation and, as such, involve simultaneous triplanar movement in all the joints of the rearfoot complex.

Key *Concepts*

The rearfoot complex refers to the functionally interdependent ankle, subtalar and midtarsal joints. The rearfoot complex facilitates pronation and supination

Arches of the feet

In combination with the ligaments and muscles in the foot, the tarsals and metatarsals are normally arranged in the form of an architectural vault, i.e. arched longitudinally (anteroposteriorly) and transversely (mediolaterally), convex on the dorsal aspect and concave on the plantar aspect (Figure 7.9A). The medial side of the longitudinal arch is formed by the calcaneus, talus, navicular, medial cuneiform and first metatarsal. The lateral side of the longitudinal arch, which is slightly shorter and much lower than the medial side, is formed by the calcaneus, cuboid and fifth metatarsal. The transverse arch is formed by the anterior five tarsals and the metatarsals. The transverse arch is largely due to the middle and lateral cuneiforms (which articulate respectively with the bases of the second and third metatarsals to form the second and third tarsometatarsal joints) and the bases of the second and third metatarsals, which are all wedge

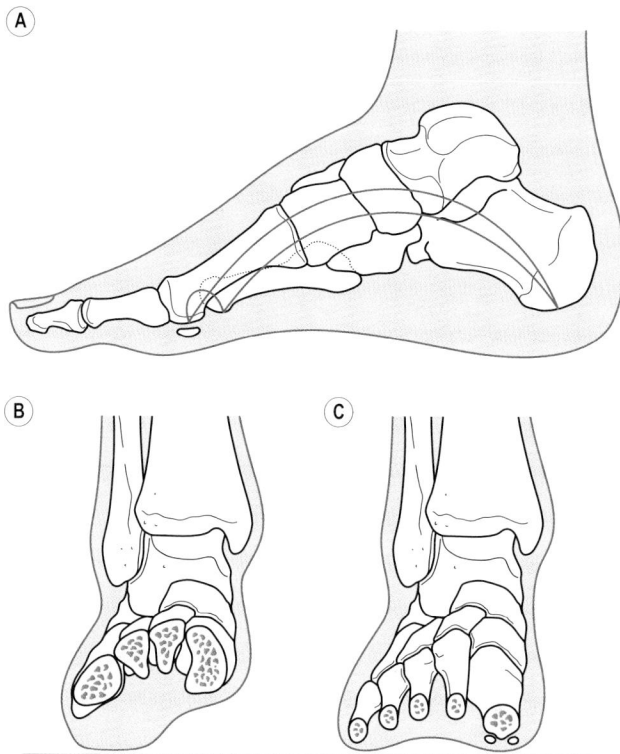

Figure 7.9 (A) Medial aspect of the right foot showing the lateral and medial parts of the longitudinal arch and the transverse arch at the heads of the metatarsals. (B) Anterior aspect of the cuneiforms and the cuboid of the right foot showing the transverse arch. (C) Anterior aspect of the heads of the metatarsals of the right foot showing the transverse arch.

shaped inferiorly in coronal section (see **Figure 7.9B**). Whereas the transverse arch is most noticeable in the region of the cuneiforms, it may also be apparent when non-weightbearing in the vertical plane through the heads of the metatarsals (see **Figure 7.9C**); weightbearing tends to flatten the transverse arch at the heads of the metatarsals.

The arches are maintained by ligaments (passive support mechanisms) and muscles (active support mechanisms). Whereas the ligaments and muscles are not very elastic, they are sufficiently elastic to enable the arches to flatten slightly following contact of the foot with the ground, such as following heel-strike in walking or running, and then recoil (restore their normal shape) following the impact. Consequently, the arches function like springs in order to help cushion impacts with the ground.

Passive arch support mechanisms

The ligaments on the plantar aspect of the foot are very strong and can normally maintain the arches of the foot in upright posture in the absence of assistance from muscles (Hicks 1961; Kitaoka et al 1997a). The main ligaments that support the arches of the foot are as follows:

1. The deep plantar calcaneocuboid ligament, also referred to as the short plantar ligament, runs from the anterior tubercle of the calcaneus to the plantar surface of the cuboid posterior to the groove for the tendon of the peroneus longus (Figures 7.10 and 7.11A and B). This ligament supports the calcaneocuboid part of the midtarsal joint.

2. The superficial plantar calcaneocuboid ligament, also referred to as the long plantar ligament, runs from the plantar surface of the calcaneus between the posterior and anterior tubercles to the plantar surface of the cuboid anterior to the groove for the tendon of the peroneus longus and to the bases of the second to fifth metatarsals (see Figures 7.10 and 7.11A and C). This ligament supports the

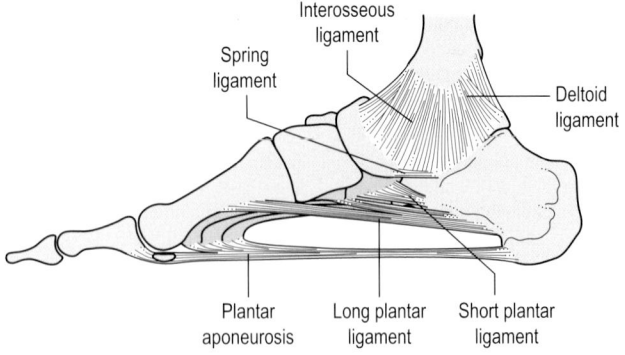

Figure 7.10 Medial aspect of the right foot showing the main arch support ligaments.

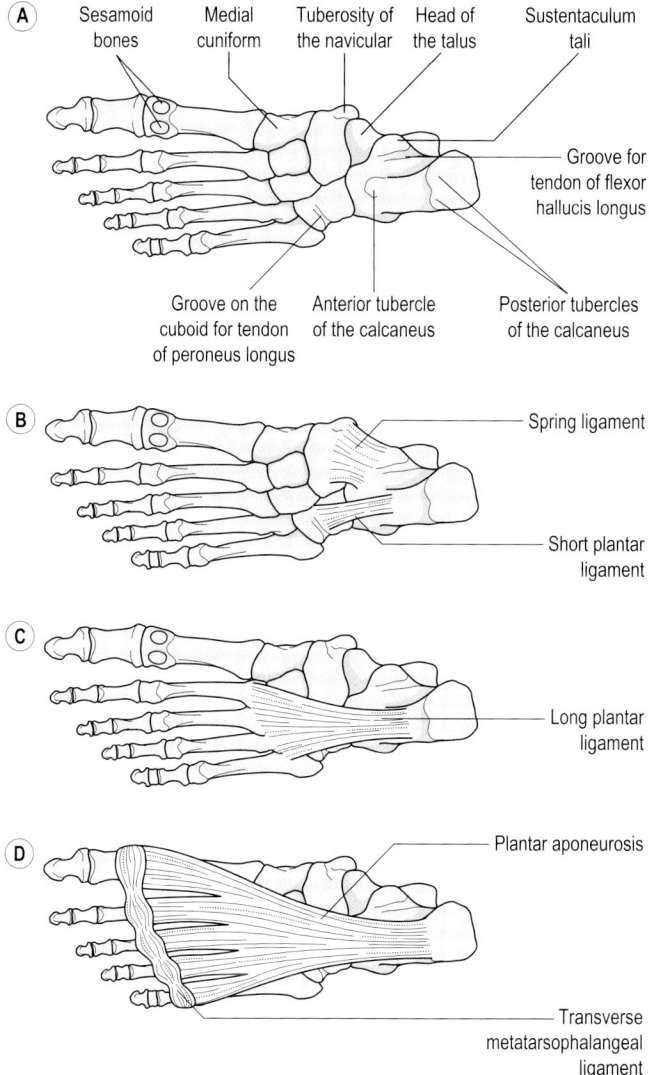

Figure 7.11 (A–D) Plantar aspect of the right foot.

calcaneocuboid part of the midtarsal joint and the lateral four tarsometatarsal joints.

3. The plantar calcaneonavicular ligament, also referred to as the spring ligament, runs from the anterior–inferior aspect of the sustentacu- lum tali (of the calcaneus) to the plantar surface of the navicular (see **Figures 7.10 and 7.11A and B**). The middle band of the deltoid ligament is attached to the medial border of the plantar calcaneonavicular liga- ment (see **Figure 7.5**). The plantar calcaneonavicular ligament supports the medial part of the subtalar joint (anterior synovial part) and the talonavicular part of the midtarsal joint.

4. The interosseous talocalcanean ligament is the syndesmosis part of the subtalar joint, described earlier.

5. The deltoid ligament (medial collateral ligament), described earlier, supports the medial aspects of the ankle and subtalar joints (see **Figure 7.5**).

6. The plantar aponeurosis is a broad fan-shaped band that spans the whole of the tarsus and metatarsus from the posterior tubercles of the calcaneus to the bases of the proximal (first) phalanges (see **Figures 7.10 and 7.11A and D**). Just anterior to the tarsometatarsal joints, the plantar aponeurosis splits into five separate bands, one to each toe. As each band passes the plantar surface of the corresponding metatarsophalangeal joint, it splits into a superficial stratum (layer) and a deep stratum. The superficial stratum attaches to the skin of the transverse sulcus, which separates the toes from the sole. The deep stratum divides into two slips, which attach, one medially and one laterally, onto the proximal plantar surface of the base of the proximal phalanx of the corresponding toe, thus forming an arch for passage of the tendon of the flexor hallucis longus (first toe) or correspond- ing tendon of the flexor digitorum longus (second to fifth toes) to the distal phalanges (Williams et al 1995). The medial and lateral slips of the plantar aponeurosis to the proximal phalanx of the hallux (first toe or great toe) merge with the tendons of the medial and lateral parts of the flexor hallucis brevis. Each tendon contains a sesamoid bone that forms a synovial joint with the plantar aspect of the head of the first metatarsal. The plantar parts of the capsules of the metatarsophalan- geal joints are thickened and referred to as plantar plates or plantar pads (Briggs 2005). The plantar plates are connected in series by deep transverse intermetatarsal ligaments and by a superficial continuous transverse metatarsophalangeal ligament (see **Figure 7.11D**). The plantar aponeurosis band to each toe merges with the corresponding plantar plate and adjoining section of the metatarsophalangeal ligament.

Mechanically, the plantar ligaments support the arches of the feet in two ways, as a beam and as a true arch (or truss) (Hicks 1961). **Figure 7.12A** shows the type of strain experienced by a loaded beam, i.e. compression strain on the upper surface and tension strain on the lower surface. This is similar to the strain on the tarsals and metatarsals by the type of arch support provided by the long plantar ligament, short plantar ligament, spring ligament, interosseous talocalcanean ligament and deltoid ligament (see **Figures 7.10 and 7.12B**). The strain on a true arch is different to that on a beam. In a true arch, the ends of the arch must move further apart if it is to become flatter and the strain on the segments of a true arch is basically compression between the segments (see **Figure 7.12C**). This is similar to the strain on the tarsals and metatarsals by the type of arch support provided by the plantar aponeurosis (see **Figure 7.12D**).

Active arch support mechanisms

The passive ligamentous beam and true arch support mechanisms are normally assisted by the muscles of the lower leg and foot. In relation

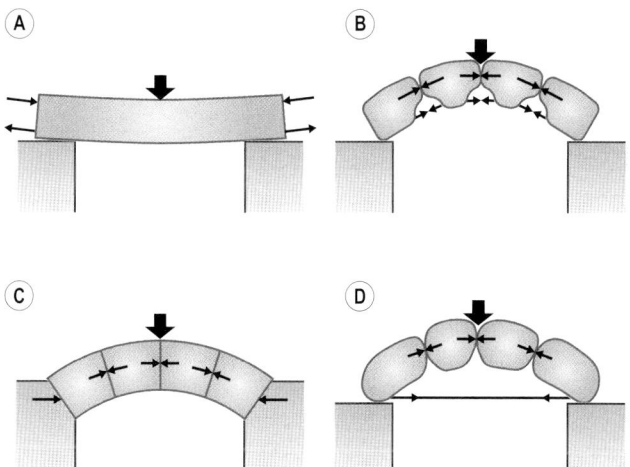

Figure 7.12 Beam and truss arch support mechanisms. (A) Strain on a horizontal beam when vertically loaded. (B) Strain on the bones of the foot and beam support mechanism when the foot is vertically loaded. (C) Strain on the components of a true arch (truss) when vertically loaded. (D) Strain on the bones of the foot and true arch support mechanism when the foot is vertically loaded.

to arch support, muscles that are located entirely (have their origins and insertions) within the foot are referred to as intrinsic muscles. Muscles that have their origins in the lower leg and insertions in the foot, i.e. cross the ankle joint, are referred to as extrinsic muscles.

The effect that a particular muscle has on the arches, i.e. tendency to raise or flatten, depends upon the tendency of the muscle to:

1. Plantar flex or dorsiflex the intertarsal, tarsometatarsal and metatarsophalangeal joints. Plantar flexion of any of these joints will tend to raise the arches and reduce the strain on the plantar ligaments. Dorsiflexion of any of the joints will tend to flatten the arches and increase the strain on the plantar ligaments.

2. Increase or decrease the ankle joint reaction force. In weightbearing, the weight of the body is transmitted to the feet via the ankle joints. Consequently, the effect of a particular weightbearing activity (standing, walking, running, hopping, jumping, etc) on the foot arches is determined by the magnitude of the ankle joint reaction forces; the greater the ankle joint reaction forces, the greater the tendency to flatten the arches, and vice versa. As demonstrated in Chapter 1, the further forward the line of action of body weight in relation to the ankle joint, the greater the magnitude of the ankle joint reaction force and, therefore, the greater the tendency to flatten the arches.

On the basis of these criteria, Hicks (1961) classified all of the intrinsic and extrinsic muscles into four groups (**Table 7.2**):

- Direct arch raiser, i.e. a tendency to plantar flex one or more of the intertarsal, tarsometatarsal and metatarsophalangeal joints.
- Indirect arch raiser, i.e. a tendency to shift body weight backward (towards the ankle joint), which tends to reduce the magnitude of the ankle joint reaction force and, therefore, reduce arch flattening.
- Direct arch flattener, i.e. a tendency to dorsiflex one or more of the intertarsal, tarsometatarsal and metatarsophalangeal joints.
- Indirect arch flattener, i.e. a tendency to shift body weight forward (in front of the ankle joint), which tends to increase the magnitude of the ankle joint reaction force and, therefore, increase arch flattening.

The lines of action of all of the plantar intrinsic muscles are below the axes of plantar flexion-dorsiflexion of one or more of the intertarsal, tarsometatarsal and metatarsophalangeal joints. Consequently, all of the plantar intrinsic muscles are plantar flexors and are classified as direct arch raisers. The effect of the extrinsic muscles on the arches depends on their lines of action in relation to the ankle joint and on their insertions (in relation to the intertarsal, tarsometatarsal and metatarsophalangeal joints). The origins,

Table 7.2 Classification of the intrinsic and extrinsic muscles of the foot in terms of their tendency to affect the arches of the feet

	Direct raise	Indirect raise	Direct flatten	Indirect flatten
Intrinsic muscles				
First layer (superficial)				
Abductor hallucis	√			
Flexor digitorum brevis	√			
Abductor digiti minimi	√			
Second layer				
Flexor accessorius	√			
Lumbicals	√			
Third layer				
Flexor hallucis brevis	√			
Adductor hallucis	√			
Flexor digiti minimi brevis	√			
Fourth layer				
Interossei	√			
Extrinsic muscles				
Flexor hallucis longus	√			√
Flexor digitorum longus	√			√
Peroneus brevis	√			√
Peroneus longus	√			√
Tibialis posterior	√			√
Extensor hallucis longus		√	√	
Extensor digitorum longus		√	√	
Tibialis anterior		√	√	
Gastrocnemius				√
Soleus				√

insertions and actions of the intrinsic and extrinsic muscles are listed in Table 7.3. The locations of the muscles are shown in Figures 7.13 and 7.14. The origins and insertions of the muscles reflect the complex relationship between the plantar ligaments and the muscles. Some of the origins and insertions of the muscles merge with plantar ligaments, especially the plantar aponeurosis. Consequently, the muscles provide arch support directly (as individual muscles) and indirectly as tighteners of plantar ligaments.

Table 7.3 Origins, insertions and actions of the intrinsic and extrinsic muscles of the foot

Muscles	Origin	Insertion	Action
Intrinsic plantar muscles			
First layer (superficial)			
Abductor hallucis (AbH)	Medial tubercle of the calacaneus and posterodorsal aspect of the plantar aponeurosis	Medial aspect of the base of the proximal phalanx of the hallux	Abduction of the hallux
Flexor digitorum brevis (FDB)	Medial and lateral tubercles of the calacaneus and posterodorsal aspect of the plantar aponeurosis	By 4 tendons, one to each of the lateral 4 toes. Each tendon splits into two branches that attach onto the sides of the shaft of the middle phalanx	Flexion of the lateral 4 toes
Abductor digiti minimi (AbDM)	Lateral tubercle of the calacaneus and posterodorsal aspect of the plantar aponeurosis	Via groove on plantar aspect of base of 5th metatarsal to lateral aspect of base of the proximal phalanx of the 5th toe	Abduction of the 5th toe
Second layer			
Flexor digitorum accessorius (FDA) (accessory to the FDL)	Medial head: line along the calcaneus inferior to the groove for the flexor hallucis longus. Lateral head: lateral tubercle of the calacaneus and posterodorsal aspect of the plantar aponeurosis	Lateral border of the tendon of the flexor digitorum longus before it splits into 4 branches	Assists the flexor digitorum longus in plantar flexion of the lateral 4 toes
Lumbricals (accessory to the FDL)	1st lumbrical: from medial aspect of FDL tendon to 2nd toe.	1st: medial aspect of the base of the proximal phalanx of the 2nd toe.	Flexion of the lateral 4 toes

	2nd lumbrical: from medial aspect of FDL tendon to 3rd toe and lateral aspect of FDL tendon to 2nd toe.	2nd: medial aspect of the base of the proximal phalanx of the 3rd toe.
	3rd lumbrical: from medial aspect of FDL tendon to 4th toe and lateral aspect of FDL tendon to 3rd toe.	3rd: medial aspect of the base of the proximal phalanx of the 4th toe.
	4th lumbrical: from medial aspect of FDL tendon to 5th toe and lateral aspect of FDL tendon to 4th toe	4th: medial aspect of the base of the proximal phalanx of the 5th toe
Third layer		
Adductor hallucis (AdH)	Oblique head: bases of the 2nd, 3rd and 4th metatarsals and sheath of the peroneus longus tendon, all merged with the plantar aponeurosis.	Oblique head: blends with the attachment of the tendon of the lateral head of the flexor hallucis brevis to the lateral aspect of the capsule of the metatarsophalangeal joint of the hallux.	Adduction of the hallux
	Transverse head: transverse metatarsophalangeal ligament over the 3rd, 4th and 5th toes	Transverse head: lateral aspect of the capsule of the metatarsophalangeal joint of the hallux and the lateral aspect of the base of the proximal phalanx of the hallux	
Flexor hallucis brevis (FHB)	Medial head: from the lateral aspect of the tendon of the tibialis posterior lateral to the tuberosity of the navicular.	Medial head: the plantar/medial aspect of the base of the proximal phalanx of the hallux.	Flexion of the hallux

(Continued)

Table 7.3 Continued

Muscles	Origin	Insertion	Action
	Lateral head: medial aspect of the cuboid distal to the groove of the peroneus longus tendon	Lateral head: the plantar/lateral aspect of the base of the proximal phalanx of the hallux	
Flexor digiti minimi brevis (FDMB)	Medial aspect of the base of the 5th metatarsal and sheath of the peroneus longus tendon	Lateral aspect of the base of the proximal phalanx of the 5th metatarsal	Abduction and flexion of the 5th toe
Fourth layer (deep)			
Dorsal interossei (DI)	4 bi-pennate muscles that arise from the superior-lateral and superior-medial aspects of the shafts of the metatarsals.		Analogous to the actions of the abductor hallucis and abductor digiti minimi
	1st interosseus: from 1st and 2nd metatarsals	Medial aspect of base of proximal phalanx of the 2nd toe	Abduction/flexion of 2nd metatarsophalangeal joint (medial abduction with respect to long axis of foot: along second metatarsal)
	2nd interosseus: from 2nd and 3rd metatarsals	Lateral aspect of base of proximal phalanx of the 2nd toe	Abduction/flexion of 2nd metatarsophalangeal joint (lateral abduction with respect to long axis of foot)
	3rd interosseus: from 3rd and 4th metatarsals	Lateral aspect of base of proximal phalanx of the 3rd toe	Abduction/flexion of 3rd metatarsophalangeal joint
	4th interosseus: from 4th and 5th metatarsals	Lateral aspect of base of proximal phalanx of the 4th toe	Abduction/flexion of 4th metatarsophalangeal joint

Plantar interossei (PI)	3 non-pennate muscles that arise from the inferior-medial aspects of the shafts of the 3rd (1st interosseus), 4th (2nd interosseus) and 5th (3rd interosseus) metatarsals	1st: medial aspect of base of proximal phalanx of the 3rd toe 2nd: medial aspect of base of proximal phalanx of the 4th toe 3rd: medial aspect of base of proximal phalanx of the 5th toe	Analogous to the action of the adductor hallucis Adduction of 3rd, 4th and 5th metatarsophalangeal joints
Intrinsic dorsal muscles			
Extensor digitorum brevis (EDB)	Anterior-superior aspect of the calacaneus in front of the sinus tarsi. Anterior aspect of the talocalcanean interosseous ligament	The medial part of the muscle is usually distinct from the lateral part. The tendon of the medial part is attached to the superior aspect of the base of the proximal phalanx of the hallux; the medial part of the muscle is sometimes referred to as the extensor hallucis brevis (EHB). The lateral part of the muscle splits into three tendons that attach onto the lateral aspects of the tendons of the extensor digitorum longus to the 2nd, 3rd and 4th toes at the metatarsophalangeal joints	Extension of the toes
Extrinsic plantar muscles			
Flexor hallucis longus (FHL)	Lower 2/3 of the posterior-medial aspect of the fibula	Under the medial malleolus to attach onto the inferior aspect of the base of distal phalanx of the hallux	Plantar flexion of the hallux. Plantar flexion and inversion of the foot

(Continued)

Table 7.3 Continued

Muscles	Origin	Insertion	Action
Flexor digitorum longus (FDL)	Lower 2/3 of the posterior aspect of the tibia	Under the medial malleolus. The tendon splits into 4 branches, one to each of the lateral 4 toes. The tendons attach onto the bases of distal phalanges of the lateral 4 toes	Plantar flexion of the lateral 4 toes. Plantar flexion and inversion of the foot
Tibialis posterior (TP)	Posterior aspect of the upper ½ of the interosseus membrane and adjoining borders of the tibia and fibula	Under the medial malleolus. The tendon slits into 6 branches that attach onto the plantar aspects of the navicular, first cuneiform and the bases of the 2nd–5th metatarsals	Plantar flexion and inversion of the foot. Raising the longitudinal and transverse arches
Peroneus longus (PL)	Head and upper 2/3 of the lateral aspect of the fibula	Under the lateral malleolus. Tendon splits into two branches that attach onto the plantar aspects of the first cuneiform and the base of the 1st metatarsal	Plantar flexion and eversion of the foot. Raising the longitudinal and transverse arches
Peroneus brevis (PB)	Lower 2/3 of the lateral aspect of the fibula	Under the lateral malleolus to attach onto the tuberosity of the 5th metatarsal	Plantar flexion and eversion of the foot. Raising the lateral aspect of the longitudinal arch
Extrinsic dorsal muscles			
Tibialis anterior (TA)	Upper 2/3 of the lateral aspect of the tibia	Anterior to the ankle joint via the medial aspect of the foot to attach onto the plantar aspects of the first cuneiform and the base of the 1st metatarsal	Dorsiflexion and inversion of the foot

Extensor hallucis longus (EHL)	Middle 2/3 of the anterior-medial aspect of the fibula	Anterior to the ankle joint to attach onto the superior aspect of the base of distal phalanx of the hallux	Dorsiflexion of the hallux. Dorsiflexion of the foot
Extensor digitorum longus (EDL)	Anterior-lateral aspect of the lateral condyle of the tibia and the upper 2/3 of the anterior aspect of the fibula	Anterior to the ankle joint. The tendon splits into 4 branches that attach onto the superior aspects of the bases of the middle and distal phalanges of the lateral 4 toes	Dorsiflexion of the lateral 4 toes. Dorsiflexion of the foot
Extrinsic calf muscles			
Gastrocnemius	Medial head: superior aspect of the medial condyle of the femur. Lateral head: superior aspect of the lateral condyle of the femur	The two heads attach onto the upper Achilles tendon, which is inserted onto the upper half of the posterior aspect of the calcaneus	Plantar flexion of the foot. Flexion of the knee
Soleus	Upper 1/3 of the fibula and adjoining soleal line on upper 1/3 of the tibia	Anterior-superior half of the Achilles tendon, which is inserted onto the upper half of the posterior aspect of the calcaneus	Plantar flexion of the foot

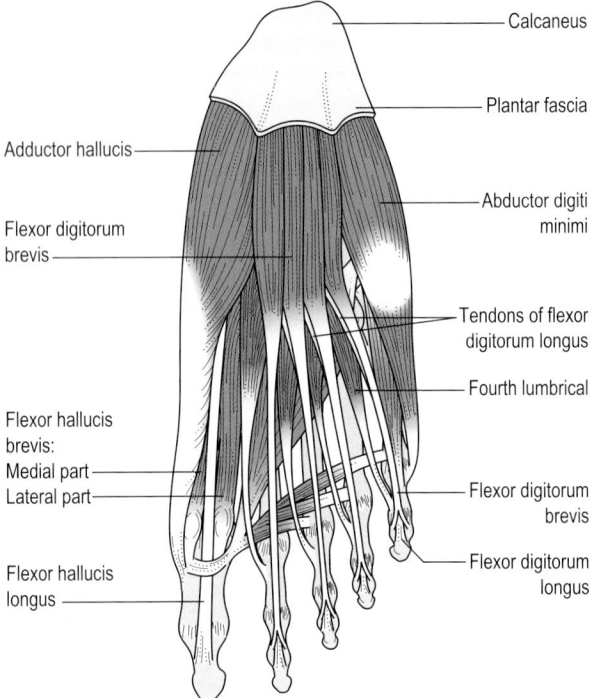

Calcaneus

Plantar fascia

Adductor hallucis

Abductor digiti minimi

Flexor digitorum brevis

Tendons of flexor digitorum longus

Fourth lumbrical

Flexor hallucis brevis:
Medial part
Lateral part

Flexor digitorum brevis

Flexor digitorum longus

Flexor hallucis longus

Figure 7.13 Intrinsic muscles of the right foot. (A) Plantar layers 1 and 2.

Key Concepts

The tarsals and metatarsals are normally arranged in the form of an architectural vault with longitudinal and transverse arches maintained by passive (ligaments) and active (muscles) mechanisms

Interaction of the arch support mechanisms

Whereas it is generally accepted that the passive (beam and true arch) and active (muscle) mechanisms both contribute significantly to arch support (Norkin & Levangie 1992), the relative contribution of the mechanisms in different weightbearing activities has yet to be determined. This lack of information reflects the difficulty of measuring the forces in the ligaments and muscles in vivo. Most studies of the arch support mechanisms have

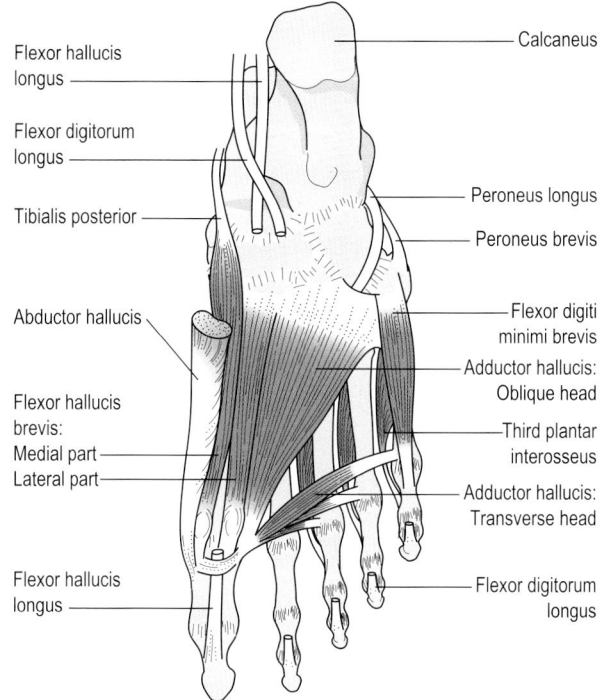

Flexor hallucis longus

Flexor digitorum longus

Tibialis posterior

Abductor hallucis

Flexor hallucis brevis:
Medial part
Lateral part

Flexor hallucis longus

Calcaneus

Peroneus longus

Peroneus brevis

Flexor digiti minimi brevis

Adductor hallucis: Oblique head

Third plantar interosseus

Adductor hallucis: Transverse head

Flexor digitorum longus

Figure 7.13 (continued) (B) Plantar layers 3 and 4.

been based on cadavers. For example, Kitaoka et al (1997a) investigated the role of the plantar ligaments in the stability of the longitudinal arches of the feet under normal loading (upright standing posture) using 19 cadaver specimens (mean age 71 years, range 20–89 years). It was found that sectioning all of the main plantar ligaments (long plantar ligament, short plantar ligament, spring ligament, interosseous talocalcanean ligament, plantar aponeurosis, deltoid ligament) resulted in complete collapse of the longitudinal arch. The arch did not collapse after sectioning any single ligament, but progessive collapse did occur when the ligaments were sectioned consecutively. The effect of sectioning individual ligaments on the degree of arch collapse (reflected in dorsiflexion of the intertarsal and tarsometatarsal joints) varied considerably among specimens, which suggested that the contribution of each ligament to arch stability varies among individuals. This is, perhaps, not surprising considering

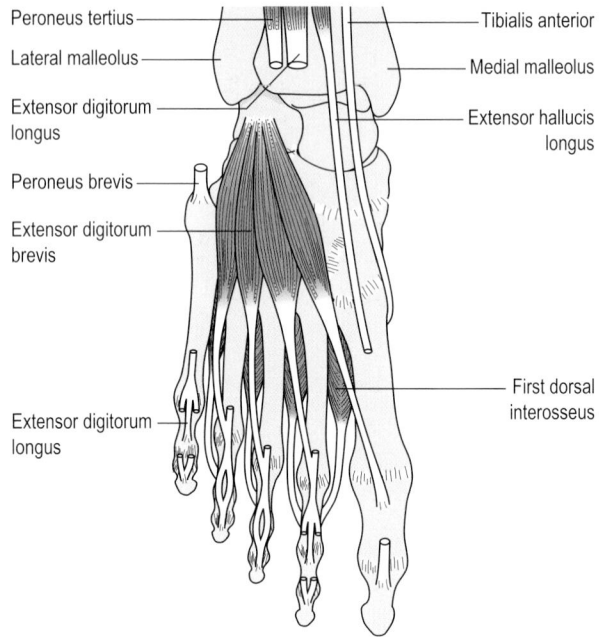

Peroneus tertius

Lateral malleolus

Extensor digitorum longus

Peroneus brevis

Extensor digitorum brevis

Extensor digitorum longus

Tibialis anterior

Medial malleolus

Extensor hallucis longus

First dorsal interosseus

Figure 7.13 (continued) (C) Dorsal.

the variation in the size, shape and alignment of the bones of the feet in normal healthy individuals (Åström & Arvidson 1995). In another cadaver study, Crary et al (2003) reported that sectioning the plantar aponeurosis resulted in a significant increase in strain on the spring ligament and long plantar ligament, i.e. the absence of the true arch mechanism resulted in a significant increase in strain on the beam mechanism.

Whereas cadaver studies provide useful information on the passive arch support mechanisms, cadaver specimens are difficult to obtain in large numbers and the results of cadaver studies do not provide information on the contribution of active support mechanisms or the relative contribution of the passive and active mechanisms. Recent studies have tried to address these problems by using detailed three-dimensional mathematical (finite element) models (Salathe & Arangio 2002; Cheung et al 2004). Salathe & Arangio (2002) investigated the effect of the extrinsic muscles (as in **Table 7.3**) on the forces in the plantar ligaments

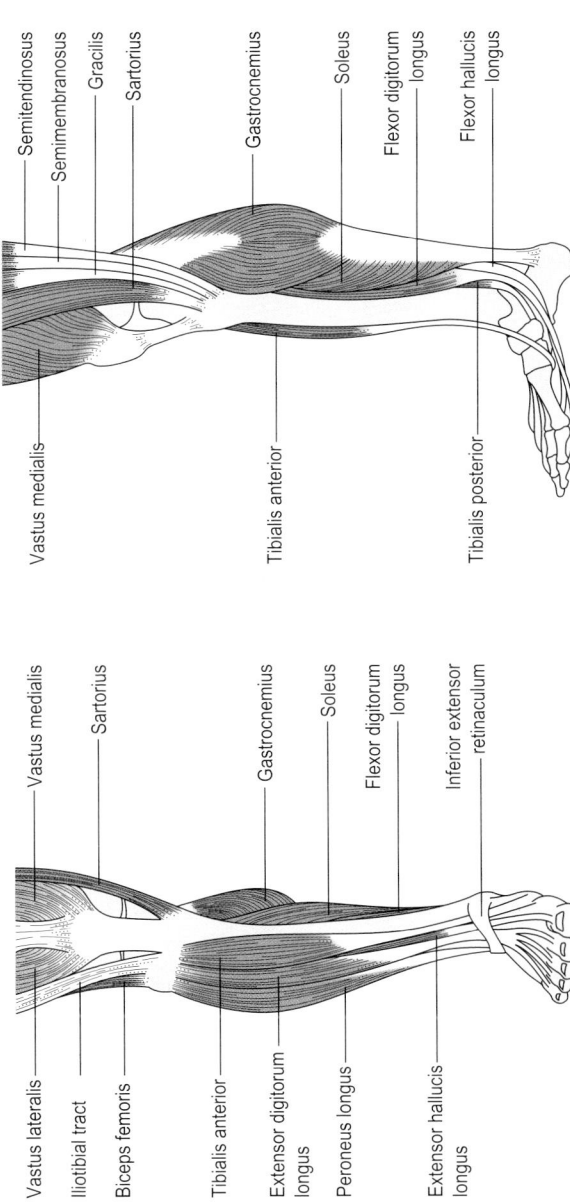

Figure 7.14 Extrinsic muscles of the right lower leg and foot. (A) Anterior. (B) Medial.

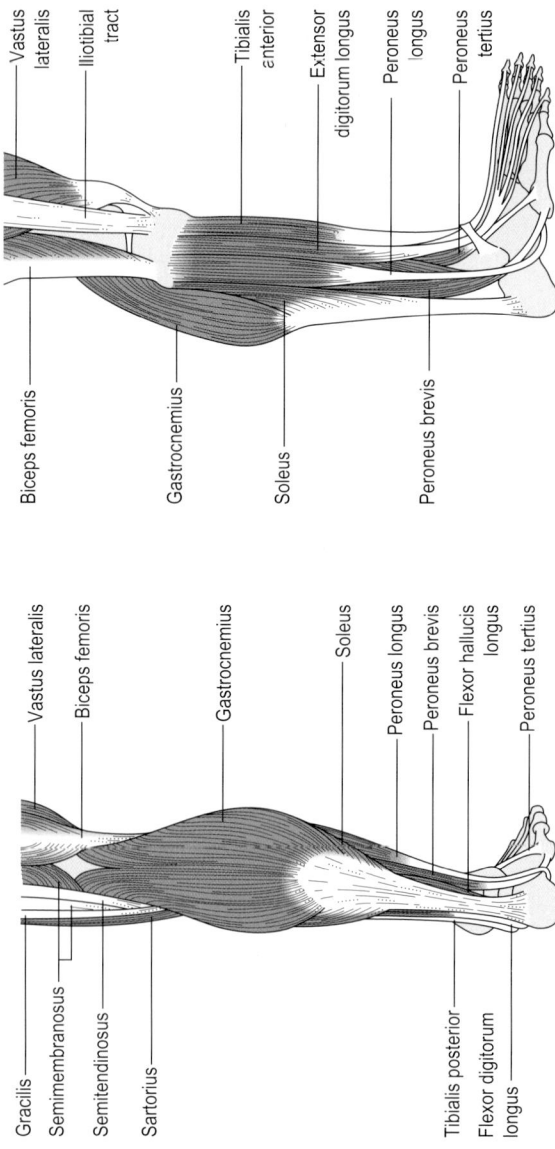

Figure 7.14 (continued) (C) Posterior. (D) Lateral.

during upright standing with the line of action of body weight in two positions, 3 cm and 17 cm in front of the ankle joints. The results indicated that the absence of extrinsic muscle support increased the forces in the long plantar ligament by 7.1% and 14.3% and in the plantar aponeurosis by 9.5% and 19.8% in the 3 cm and 17 cm conditions, respectively. The authors acknowledged that the intrinsic muscles are likely to have a similar if not greater additional effect, but the intrinsic muscles were not included in the model, presumably due to the difficulty in modelling these muscles.

Podiatric biomechanics

There is general agreement amongst podiatrists that most disorders of the foot are caused by chronic overload (prolonged excessive loading on parts of the skin and musculoskeletal components) due to abnormal foot function during weightbearing activity (Harradine et al 2006). Podiatric biomechanics is the branch of biomechanics concerned with the study of the forces that act on and within the foot and lower leg during weightbearing activity and the effects of the forces on the structure and function of the foot. Consequently, podiatric biomechanics underlies the assessment and treatment of most foot disorders treated by podiatrists (Payne 1998). At present, there is no universally agreed theory of foot function during weightbearing activity and, consequently, no universal approach to assessment and treatment of foot disorders (Harradine et al 2003). The absence of a universally agreed theory would appear to be due, at least in part, to the difficulty of measuring the loads on the components of the foot during gait. Whereas it is possible, through the use of force plates and pressure platforms/mats/insoles, to measure external forces and pressures on the foot, current technology does not allow easy measurement of internal forces and stresses on muscles, tendons, ligaments and articular surfaces of joints during gait (Sutherland 2001).

Notwithstanding the lack of a universally agreed theory of foot function during weightbearing activity, the function of the foot during gait (walking) is to transmit loads between the lower leg and the ground in an effective and efficient manner (Gage 1990; Kuo 2007). Effective load transmission results in the maintenance of controlled walking (forward movement of the body in an upright posture by means of alternate steps of the left and right legs). Efficient load transmission conserves energy and reduces energy expenditure. In order to achieve these objectives, the foot must function as both shock absorber and propulsion system in different phases of the gait cycle.

Key Concepts

> Podiatric biomechanics is the study of the forces that act on and within the foot and lower leg during weightbearing, and the effects of the forces on the structure and function of the foot

Gait cycle

In walking, a step is defined as the movement of the body from contact of one foot with the ground to contact of the other foot with the ground. A stride is defined as the movement of the body during two successive steps. The gait cycle refers to the movement of the body during a single stride. The gait cycle of the right leg begins with right heel-strike (contact of the ground with the right heel), which initiates the stance phase of the right leg, i.e. the period of the gait cycle when the right leg is in contact with the ground. The first part of the stance phase is a period of double support, i.e. when both feet are in contact with the ground (Figure 7.15). This period of double support lasts for approximately 10% of the cycle at which point the left foot leaves the ground (referred to as toe-off) and the left leg swings forward. During the swing phase of the left leg the right leg

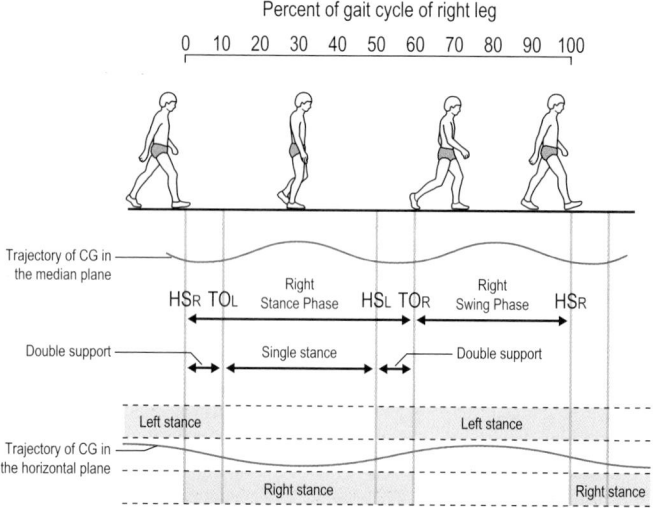

Figure 7.15 Walking gait cycle. CG, centre of gravity. HS, heel-strike. TO, toe-off.

supports the body on its own; this period lasts for approximately 40% of the cycle and is referred to as the single-support phase of the right leg. At the end of the swing phase of the left leg the left foot contacts the ground and another period of double-support ensues. At approximately 60% of the cycle the right foot leaves the ground to begin its swing phase while the left leg experiences a period of single support. The cycle is completed by heel-strike of the right foot.

Trajectory of the centre of gravity

During walking, the movement of the body as a whole is reflected in the movement of the whole body centre of gravity, which tends to follow a fairly smooth up-down side-to-side trajectory. When viewed from the side, as shown in the upper part of Figure 7.15, the centre of gravity moves up and down twice during each gait cycle with the low points of the trajectory occurring close to the mid-points of the double-support phases and the high points of the trajectory occurring close to the mid-points of the single-support phases. When viewed from overhead, as shown in the lower part of Figure 7.15, the trajectory of the centre of gravity follows the support phases, moving right during the period from the mid-point of single-support of the left leg to the mid-point of single-support of the right leg and moving left during the period from the mid-point of single-support of the right leg to the mid-point of single-support of the left leg. As shown in Figure 7.16, the vertical excursion (up-down range of motion) of the centre of gravity during gait increases with increasing speed and ranges from approximately 2.7 cm at 0.7 m/s (slow walk: 1.57 mph) to approximately 4.8 cm at 1.6 m/s (moderate walking speed: 3.58 mph) (Orendurff et al 2004). In contrast, the mediolateral excursion (side-to-side range of motion) of the centre of gravity during gait decreases with increasing speed and ranges from approximately 7.0 cm at 0.7 m/s to approximately 3.8 cm at 1.6 m/s.

Components of the ground reaction force

Figure 7.17 shows the ground reaction force (F) and the anteroposterior (F_X), vertical (F_Y) and mediolateral (F_Z) components of the ground reaction force at one point in the single-support phase of the right leg. When walking straight forward, the mediolateral component is normally very small resulting in little side-to-side movement of the body. Figure 7.18 shows the anteroposterior, vertical and mediolateral components of the ground reaction force (force-time curves) exerted on each leg during the gait cycle. The movement of the centre of gravity during the gait cycle is determined by the resultant force acting on it. During a period of single-support the resultant force acting on the centre of gravity is determined

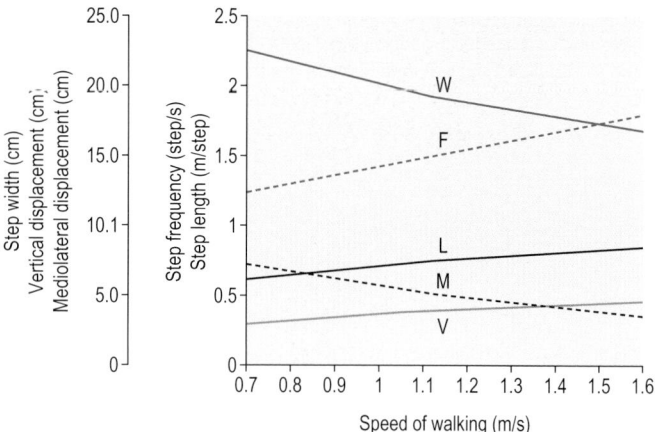

Figure 7.16 Effect of walking speed on step frequency (*F*), step length (*L*), step width (*W*), vertical displacement of the centre of gravity (*V*) and mediolateral displacement of the centre of gravity (*M*). Based on data from Orendurff MS, Segal AD, Klute GK, et al (2004) The effect of walking speed on center of mass displacement. Journal of Rehabilitation Research & Development 41(6A): 829–34.

by body weight and the ground reaction force exerted on the grounded foot. During a period of double-support the resultant force acting on the centre of gravity is determined by body weight and the ground reaction forces exerted on both feet.

The vertical component of the ground reaction force exerted on each leg is characteristically dominated by two smooth peaks, with the rise and fall of each peak taking up about half of the stance phase (see Figure 7.18). The rise and fall of the first peak roughly corresponds to the period from heel-strike to heel-off and the rise and fall of the second peak roughly corresponds to the period from heel-off to toe-off.

Like the vertical component, the anteroposterior component is normally characteristically dominated by two smooth peaks whose rise and fall correspond to the rise and fall of the two peaks of the vertical component. The resultant anteroposterior component of force acts backward from the mid-point of double-support to heel-off (a braking force), indicating deceleration of the centre of gravity, i.e. the forward speed of the body is decreased. From heel-off to the mid-point of the next double-support, the resultant anteroposterior component acts forward, indicating forward acceleration of the centre of gravity, i.e. the forward speed of the body is increased.

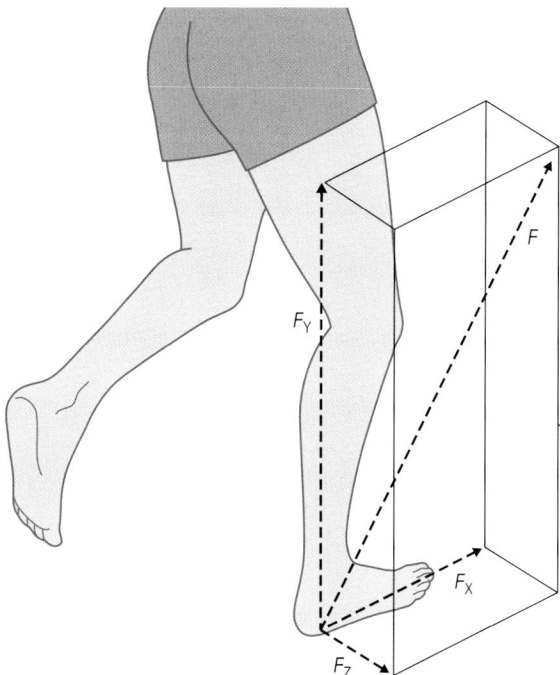

Figure 7.17 Anteroposterior (F_X), vertical (F_Y) and mediolateral (F_Z) components of the ground reaction force (F).

The resultant mediolateral component of force acting on the centre of gravity during the gait cycle acts medially during single stance and changes direction during double-support, i.e. from medial on the right foot to medial on the left foot during the period from left heel-strike to right toe-off (see **Figure 7.18**).

In addition to the characteristic smooth phases of the vertical, antero-posterior and mediolateral components of the ground reaction force on each foot, all three components are often characterized by a single or multiple transient spikes soon after heel-strike, which reflect the impact of the heel with the ground (see F_Y in **Figure 7.18**). Shock absorbing foot-wear will reduce or eliminate these transient spikes (Czerniecki 1988).

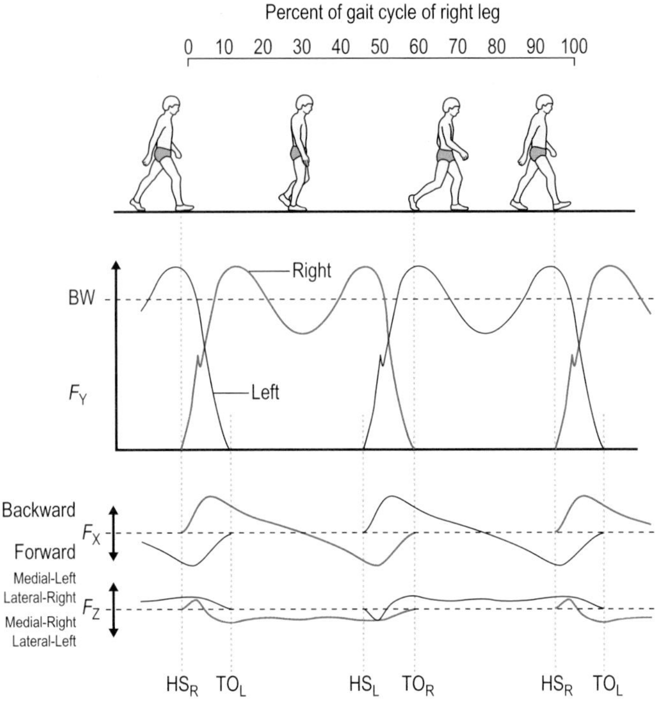

Percent of gait cycle of right leg

0   10   20   30   40   50   60   70   80   90   100

BW

F_Y

Right

Left

Backward

F_X

Forward

Medial-Left
Lateral-Right

F_Z

Medial-Right
Lateral-Left

HS$_R$ TO$_L$ HS$_L$ TO$_R$ HS$_R$ TO$_L$

Figure 7.18 Anteroposterior (F_X), vertical (F_Y) and mediolateral (F_Z) components of the ground reaction force (F) during the walking gait cycle. BW, body weight; HS, heel-strike; TO, toe-off.

Path of the centre of pressure

As indicated in **Figure 7.18** the magnitude and direction of the ground reaction force change continuously during the gait cycle. **Figure 7.19** shows the change in the resultant (F_{XY}) of the anteroposterior (F_X) and vertical (F_Y) components. Owing to the dominance of F_Y, the change in F_{XY} from heel-strike to toe-off reflects the double-peaked F_Y-time curve component as in **Figure 7.18**. F_Y always acts upward so that the progressive change in direction of F_{XY} from upward and backward at heel-strike to more-or-less vertical at heel-off to upward and forward at toe-off is largely due to the change in the direction of F_X from backward (heel-strike to heel-off) to forward (heel-off to toe-off) (see **Figure 7.18**).

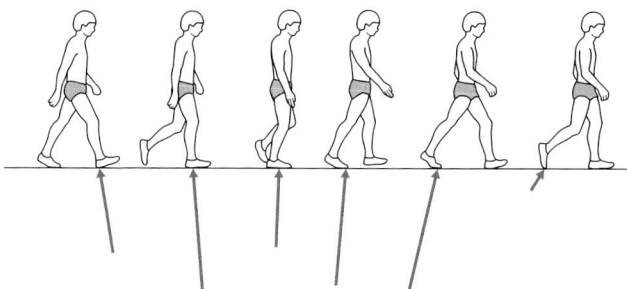

Figure 7.19 Change in the magnitude and direction of the ground reaction force on the right foot during the stance phase in gait.

During stance, the foot essentially rolls forward from heel to toe such that the contact area between foot and ground and, consequently, the centre of pressure, change continuously (Figure 7.20A to E). Hansen et al (2004) showed that in normal gait, the forward movement of the centre of pressure is similar to that of a rolling wheel with radius approximately 30% of leg length (see Figure 7.20F). They also found that this radius of curvature does not change significantly with changes in walking speed, shoe height and load carried.

The notion that in normal gait the foot and the body as a whole moves across the ground in a wheel-like manner, which reduces energy expenditure, has been expressed by several investigators over many years (e.g. Saunders et al 1953; Cavagna & Margaria 1966; McGeer 1990; Alexander 1995). Belief in the wheel-like action of the foot resulted in the design of shoes with rocker-bottom surfaces that facilitate a rolling foot action. These shoes have been used successfully (delaying/arresting the onset/progression of diseases, reducing plantar pressure, reducing energy expenditure) in the treatment of a variety of foot disorders including peripheral neuropoathy, diabetic ulcers and transmetatarsal amputation (Schaff & Cavanagh 1990; Praet & Louwerens 2003). By attaching rigid arc shapes of various radii to the bottoms of rigid walking boots, which restricted ankle movement, Adamczyk et al (2006) examined the effect of varying foot curvature on the mechanics and energy expenditure of walking. The results showed that energy expenditure (oxygen uptake) in normal gait is minimal when the effective radius of curvature of the foot is approximately 30% of leg length. As the length of the foot is also approximately 30% of leg length, it would appear that in normal gait the coordinated action of the joints of the lower limb (hip, knee, ankle, foot) result in

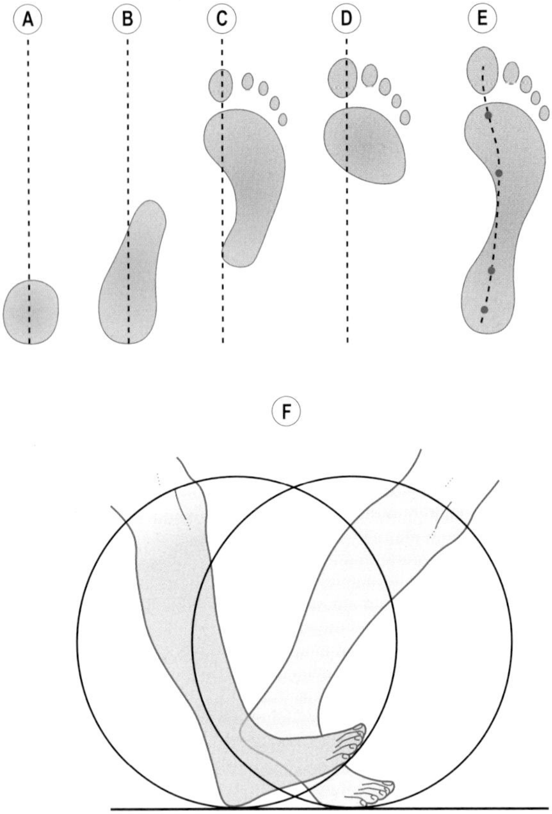

Ⓐ Ⓑ Ⓒ Ⓓ Ⓔ

Ⓕ

Figure 7.20 Movement of the centre of pressure along the plantar surface of the right foot during the stance phase in gait. Plantar contact area at (A) HS$_R$, middle of double support (B), heel-off right (C) and heel-strike left (D). (E) Path of the centre of pressure and location of the centre of pressure at the points corresponding to (A–D), respectively. (F) Normal movement of the lower leg and foot during gait.

movement of the plantar surface of the foot across the ground in a way that minimizes energy expenditure and that this mechanism is most efficient when the length of the foot is approximately 30% of leg length.

To understand how the body minimizes energy expenditure in gait, it is necessary to understand the relationship between energy expenditure and work, i.e. the transformation of chemical energy in the cells of the body to mechanical energy in the movement of the body.

Energy expenditure and work

There are a number of different forms of energy including heat, light, sound, electricity, chemical energy and various forms of mechanical energy. The total amount of energy in the universe is constant; it cannot be created or destroyed, it can only be transformed from one form to another. All interactions in nature are the result of transformation of energy from one form to another. Living organisms digest food to release nutrients that are contained in the food. Nutrients, including carbohydrates, fats, proteins, minerals and vitamins are used by the body for growth, maintenance and repair. Some of the nutrients, in particular carbohydrates and fats, are used to produce chemical energy, i.e. substances that, when required, break down into their constituent components and simultaneously release energy to maintain all of the processes that sustain life. Metabolism refers to all of the chemical reactions that take place in the body and metabolic energy expenditure, usually referred to simply as energy expenditure, refers to the amount of energy used in these chemical reactions. Metabolic rate refers to the rate of metabolic energy expenditure and is usually expressed in relation to body mass and time as, for example, energy expenditure per kilogram of body mass per minute.

Oxygen is necessary to regenerate chemical energy from its constituent components following breakdown and release of energy. As there is a direct relationship between energy expenditure and oxygen consumption, energy expenditure is usually measured in terms of oxygen uptake and often expressed in METs. One MET (metabolic equivalent) refers to resting metabolic rate, i.e. resting level of oxygen consumption. One MET varies slightly among individuals, but is approximately $3.5\,ml.kg^{-1}.min^{-1}$ of oxygen. METs are used to compare the energy costs of different activities (Kent 2006).

The majority of the energy produced from nutrients is used by the muscles to produce mechanical energy in the form of movement of the body segments. Transformation of energy into mechanical energy is referred to as work. All forms of energy are equivalent in their capacity to do work, i.e. bring about the transfer of energy from one body to another through the action of a force or forces that deform and/or change the position and/or speed of movement of the bodies.

A force does work when it moves its point of application in the direction of the force and the amount of work done is defined as the product of the force and the distance moved by the point of application of the force. For example, in drawing a bow, the archer does work on the bow by pulling on the arrow which, in turn, pulls on the bowstring (Figure 7.21).

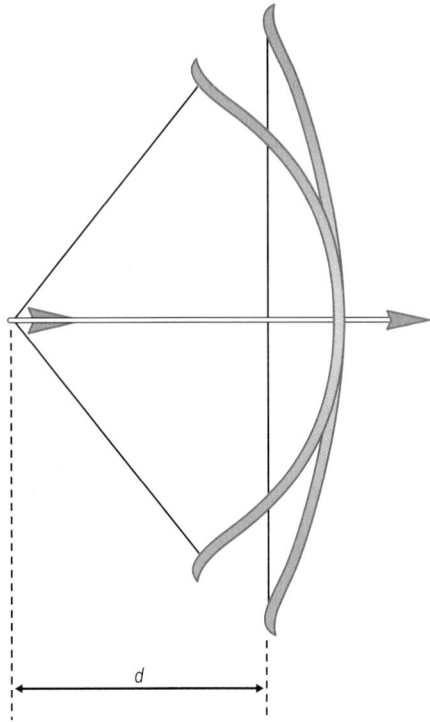

Figure 7.21 Storage of strain energy in a bow. *d*, distance that the bow is drawn.

As the bow is drawn, the work done on the bow is stored in the bow which deforms like a spring; the greater the deformation of the bow, the greater the amount of energy that is stored in it. The amount of work done by the archer on the bow is $F.d$ where F is the average force exerted on the bowstring and d is the distance that the bow is drawn back. When the arrow is released, the bow recoils and the stored energy is transformed into work on the arrow via the bowstring. The arrow separates from the bowstring with kinetic energy equivalent to the work done on it, i.e. the energy stored in the drawn bow. Kinetic energy, a form of mechanical energy, is the energy possessed by a body due to its speed of movement. An object of mass m and speed of movement v has kinetic energy equal to $m.v^2/2$. Consequently, in the bow and arrow example, $F.d = m.v^2/2$ where F = average force exerted on the arrow by the bowstring,

d = distance over which F is applied, m = mass of arrow and v = release speed of the arrow.

The energy stored in a drawn bow is referred to as strain energy, another form of mechanical energy. Many materials store strain energy in response to loading, for example, a stretched elastic band, a trampoline, a springboard in diving, a beat board in vaulting, a pole in pole vaulting and a muscle–tendon unit in an eccentric contraction. Strain energy is a form of potential energy, i.e. stored energy that, given appropriate conditions, may be used to do work.

Human movement is brought about by coordinated actions of the skeletal muscles. The muscles do work in moving the body segments. In doing work, the muscles transform chemical energy stored in the muscles into kinetic energy of the moving body segments. The stored energy in the muscles is potential energy in the form of complex chemical substances (chiefly adenosine triphosphate and creatine phosphate). When a muscle contracts isometrically (no change in length) it expends energy but does no work. When a muscle contracts concentrically (shortens) it expends energy in doing positive work, i.e. it pulls its skeletal attachments closer together. Positive work increases the mechanical energy of the body. When a muscle contracts eccentrically (lengthens) it expends energy, but in contrast to a concentric contraction, work is done on the muscle–tendon unit, i.e. it is forcibly stretched; this type of work by muscles is called negative work. Negative work decreases the mechanical energy of the body. As a muscle–tendon unit lengthens in an eccentric contraction, the elastic components of the muscle–tendon unit absorb energy in the form of strain energy. For example, when landing from a jump or vault in gymnastics, the kinetic energy of the body is transformed into strain energy in the support surface and strain energy in the muscle–tendon units that control the hip, knee and ankle joints by eccentric contraction of these muscles. When the purpose of the landing is to bring the body to rest, the strain energy in the muscle–tendon units is rapidly dissipated as heat in the muscle–tendon units and subsequently in the rest of the body and the surrounding air. However, if the landing is immediately followed by a rebound, some of the strain energy in the muscle–tendon units may be recycled in the subsequent movement in the form of work (additional to that produced by concentric action of the muscles) resulting from recoil of the elastic components of the muscle–tendon units. The use of this strain energy depends largely on the speed of changeover from eccentric to concentric muscle contraction. Generally, the faster the changeover, the smaller the proportion of strain energy dissipated as heat and, consequently, the greater the proportion available to contribute to the subsequent movement (Gregor 1993).

Key *Concepts*

Living organisms produce chemical energy from food to maintain the processes that sustain life. Metabolism refers to all of the chemical reactions that take place in the body and energy expenditure refers to the amount of energy released in these chemical reactions

Gravitational potential energy

When walking, the centre of gravity of the body moves upward and forward during the first half of single support, i.e. from toe-off to the middle of single-support, and downward and forward during the second half of single-support, i.e. from the middle of single-support to heel-strike (see **Figure 7.15**). The upward and forward movement of the centre of gravity in the first half of single-support is largely the result of positive work by the leg extensor muscles of the trailing leg during the push-off prior to toe-off. As the centre of gravity moves up, the gravitational potential energy of the body increases, i.e. the energy possessed by the body due to the height of its centre of gravity above the ground; the higher the centre of gravity, the greater the gravitational potential energy. Like kinetic energy and strain energy, gravitational potential energy is a form of mechanical energy. The gravitational potential energy possessed by the body at the middle of single-support is converted to kinetic energy during the second half of single-support, i.e. the speed of the centre of gravity increases forward and downward as the gravitational potential energy decreases. Consequently, following toe-off, the kinetic energy of the body resulting from the push-off is largely converted to gravitational potential energy as the centre of gravity moves up. The gravitational potential energy is then converted to kinetic energy as the centre of gravity moves down. As will be seen later in the chapter, the interchange of kinetic energy and gravitational potential energy during the gait cycle decreases energy expenditure and increases mechanical efficiency.

The interchange of gravitational potential energy and kinetic energy can, perhaps, be best illustrated by considering the movement of a falling object. If an object is held above ground level and then released, it will fall to the ground due to the force of its own weight. The work done on the object by the force of its own weight W when it falls a distance h is given by $W.h$. Consequently, when an object of weight W is held a distance h above the ground, it possesses gravitational potential energy equivalent to $W.h$, which may be transformed into kinetic energy if it is allowed to fall. Gravitational potential energy is usually denoted $m.g.h$ (where $W = m.g$).

Figure 7.22A shows a rubber ball held at rest at a height h_1 above the floor where the floor is the reference level ($h = 0$) for the measurement of gravitational potential energy. While it is held at rest, the ball has no kinetic energy, but its gravitational potential energy will be equal to $m.g.h_1$. If the ball is allowed to fall, its gravitational potential energy will be transformed into kinetic energy, i.e. its gravitational potential energy will decrease and its kinetic energy will increase. When the ball hits the floor,

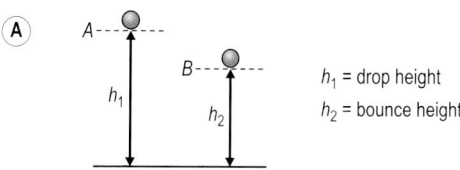

h_1 = drop height
h_2 = bounce height

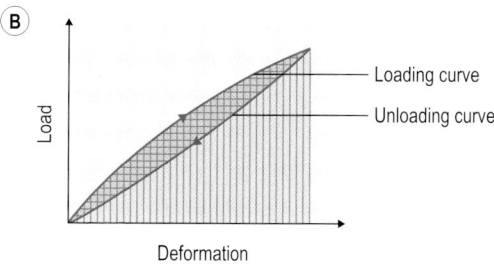

Loading curve
Unloading curve

Load

Deformation

||||| & ▦ = G_L = Strain energy absorbed by the ball during loading, i.e. the area under the loading curve

||||| = G_U = Energy returned by the ball during unloading, i.e. the area under the unloading curve

▦ = Energy dissipated during contact with the floor

The resilience R of the ball is given by: $R = \dfrac{G_U}{G_L} \times 100\%$

If G_L = 180 J and G_U = 153 J, then $R = \dfrac{153 \text{ J}}{180 \text{ J}} \times 100\% = 85\%$

Figure 7.22 Load-deformation characteristics of a bouncing ball.

its gravitational potential energy will be zero and its kinetic energy will be equal to $m.v^2/2$ where v is the velocity of the ball at impact, i.e.

$$m \cdot g \cdot h_1 = m \cdot v^2/2$$
$$v = \sqrt{(2g \cdot h_1)}$$

If $h_1 = 1\,\mathrm{m}$, then

$$v = \sqrt{(2 \times 9.81\,\mathrm{m/s}^2 \times 1\,\mathrm{m})} = 4.43\,\mathrm{m/s}$$

Hysteresis, resilience and damping

In the above example, the ball strikes the floor with kinetic energy equivalent to the gravitational potential energy it possessed at release. During contact with the floor the ball will undergo a loading phase in which it is compressed and the kinetic energy of the ball is transformed into strain energy in the compressed ball. Following the loading phase the ball undergoes an unloading phase in which it recoils and the strain energy is released as kinetic energy in the form of the upward bounce of the ball. However, the ball will not bounce as high as the point from which it was dropped. This situation is shown in **Figure 7.22A** where h_1 is the drop height and h_2 is the bounce height. As the ball is at rest at A and B, some of the energy of the ball was dissipated during contact with the floor in the form of, for example, heat and sound. The amount of energy dissipated is reflected in the load deformation curves of the ball during loading and unloading (see **Figure 7.22B**).

The amount of strain energy absorbed by the ball during loading, the area under the loading curve, is greater than the amount of energy returned during unloading, the area under the unloading curve. The loop described by the loading and unloading curves is the hysteresis loop. The area of the hysteresis loop represents the energy dissipated. The extent of hysteresis in a material is reflected in the resilience of the material, which is defined as the amount of energy returned during unloading as a percentage of the amount of energy absorbed during loading. All materials exhibit hysteresis to a certain extent and, consequently, there are no 100% resilient materials.

Figure 7.23A shows the load-deformation characteristics of highly resilient material such as ligament and tendon, and **Figure 7.23B** shows the load-deformation characteristics of low-resilience material, such as some forms of vinyl acetate foam. Damping refers to a low level of resilience; a damping material returns very little energy during unloading compared to

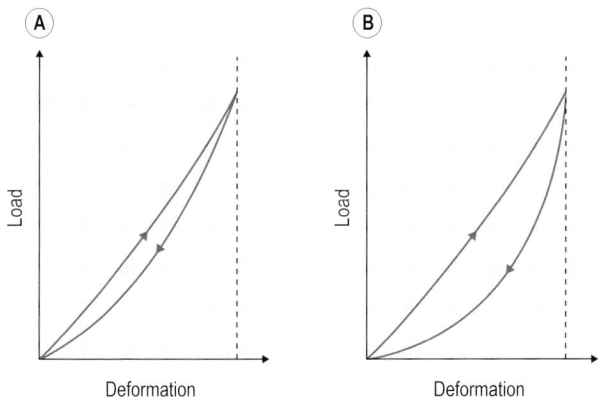

Figure 7.23 Load-deformation characteristics of material with high resilience (A) and low resilience (B).

the amount of energy that it absorbs during loading. For protection during transportation, fragile goods are usually packed in materials with good damping properties. Similarly, in walking and running, shock absorbing soles and insoles in shoes are used to protect the body from high impact loads at heel-strike. In human movement, shock absorption refers to the dissipation of the work done on the body as a result of a collision with the environment in a way that prevents high impact loads. Shock absorption systems are low-resilience energy absorption systems.

Key *Concepts*

The human body converts chemical energy into three forms of mechanical energy: kinetic energy, gravitational potential energy and strain energy

Theories of gait

As shown in **Figure 7.15**, the overall forward movement of the centre of gravity in normal gait is associated with vertical and mediolateral displacements resulting in a sinusoidal path. The sinusoidal path of the centre of gravity is the result of sinusoidal changes in the speed of the centre of gravity in forward, vertical and mediolateral directions (**Figure 7.24**).

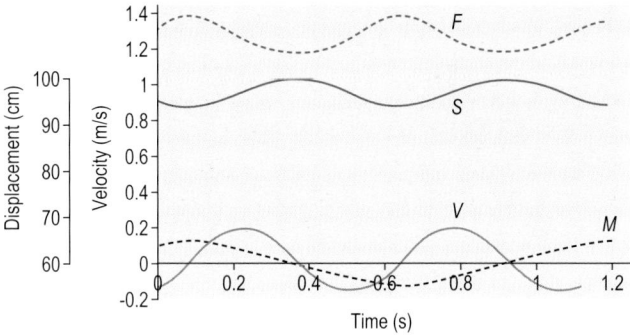

Figure 7.24 Forward (*F*), vertical (*V*) and mediolateral (*M*) components of the velocity of the centre of gravity and vertical displacement (*S*) of the centre of gravity of a healthy adult walking at 1.25 m/s during one complete gait cycle (heel-strike to heel-strike of the right foot). Based on data from Donelan JM, Kram R, Kuo AD (2002) Simultaneous positive and negative external work in human walking. Journal of Biomechanics 35:117–24.

Consequently, in normal gait, the centre of gravity experiences continuous changes in gravitational potential energy and kinetic energy. Any change in the kinetic energy or gravitational potential energy of the centre of gravity requires energy expenditure, and the greater the change in energy, the greater the energy expenditure.

Theoretically, changes in the kinetic energy and gravitational potential energy of the centre of gravity during gait and, therefore, the energy expenditure required to bring about these changes, could be reduced to zero by moving the centre of gravity in a straight line parallel to the ground (no change in gravitational potential energy) at constant speed (no change in kinetic energy). However, energy expenditure in self-propelled movement depends on the design of the moving object as well as on the changes in gravitational potential energy and kinetic energy of the centre of gravity of the object that occur during movement.

When a particular movement task is clearly defined, the machine required to carry out the task can be specifically designed for that purpose. Consequently, if the only function of the machine is to transport a mass, then a wheel is the most efficient design (**Figure 7.25A**). The centre of gravity of a wheel is at its centre so that when the wheel rolls on a flat level surface the path of the centre of gravity will be a straight line with no vertical or side-to-side deviation. Consequently, after an initial push to start the wheel rolling, there will be no change in the gravitational potential energy or kinetic energy of the wheel and, therefore, no energy expenditure will be needed to keep it rolling.

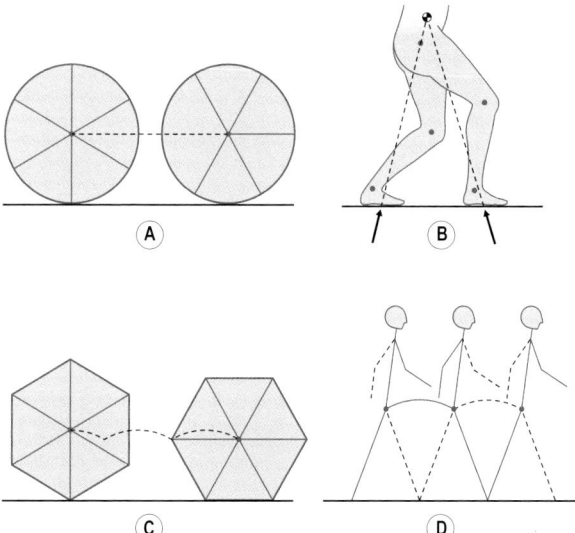

Figure 7.25 (A) Path of the centre of gravity of a rolling wheel. (B) Walking while attempting to keep the centre of gravity level. (C) Path of the centre of gravity of a hexagon rotating on its edge. (D) Compass gait.

In contrast to a wheel, the human neuromusculoskeletal system can perform many more functions than simply transporting the body, and the energy expenditure of the body in different activities reflects the constraints of its structure. In normal gait, the knee joints do not normally flex more than about 30° during the stance phases (Gard & Childress 2001). In order to move the centre of gravity of the human body in a straight line, parallel with the ground, the knee joints of both legs would need to be permanently flexed 70°–90° during the stance phases (Kuo 2007) (see Figure 7.25B). As the ground reaction forces on the feet are usually directed through the centre of gravity, the moments of the ground reaction forces about the knees, and the corresponding muscle moments about the knees, would result in a very high level of energy expenditure. It has been shown that humans find it very difficult to reduce the vertical excursion of the centre of gravity during gait by more than about half that in normal gait and that even this moderate decrease more than doubles the energy expenditure (Ortega & Farley 2005).

For a particular speed of walking, humans normally adopt a particular combination of step frequency, step length and step width that requires

least energy expenditure (Holt et al 1995; Jeng et al 1997; Bertram & Ruina 2001). The movement of the centre of gravity of the human body in normal gait is a feature of an overall movement pattern that is optimal in terms of the energy expenditure needed to transport the body at a particular speed. Energy expenditure is minimized by maximizing energy conservation through exchanges between gravitational potential energy, kinetic energy and strain energy. Not surprisingly, energy conservation is the basis of the two most well-documented theories of gait, the six determinants of gait theory and the inverted pendulum theory of gait.

The six determinants of gait theory

The simplest model of gait is that in which the legs are represented by two struts articulated at the equivalent of a hip joint, but without feet, ankles and knees. In this model, each leg has one degree of freedom, i.e. hip flexion–extension, and the path of the centre of gravity would be a succession of arcs, like stepping off distances with a pair of compasses (see **Figure 7.25C and D**). This form of gait has been referred to as compass gait (Saunders et al 1953). The energy expenditure of compass gait is reflected in the vertical excursion of the centre of gravity in each step, i.e. changes in gravitational potential energy, and abrupt changes in the speed and direction of the centre of gravity at the end of one step and the start of the next, i.e. changes in kinetic energy. The energy expenditure required to maintain compass gait is approximately double that required in normal gait (Kuo 2007).

In comparison with the one degree of freedom struts in the compass gait model, each human leg has six degrees of freedom, three at the hip (internal–external rotation, flexion–extension, abduction–adduction) and one each at the knee (flexion–extension), ankle (plantar flexion–dorsiflexion) and metatarsophalangeal (flexion–extension) joints (**Figure 7.26A**). Consequently, the displacement of the centre of gravity in gait can be influenced by movements in any of these joints. Saunders et al (1953) suggested that humans reduce the energy expenditure of gait, relative to compass gait, by coordinating the action of the hip, knee, ankle and foot joints to reduce the vertical and mediolateral displacement of the centre of gravity, and to smooth the change in speed and direction of the centre of gravity between successive steps. Specifically, Saunders et al (1953) referred to six determinants of locomotion:

1. Pelvic rotation during each step (see **Figure 7.26B**): internal–external rotation of the thigh with respect to the transverse plane. The pelvis rotates forward about a vertical axis through the hip of the stance leg in each step, i.e. the hip joint of the swing leg moves from behind

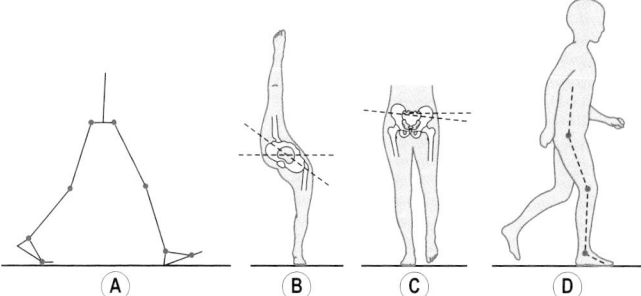

Figure 7.26 Degrees of freedom of the legs. (A) The path of the centre of gravity can be influenced by movement in the hips, knees, ankles and metatarsophalangeal joints. (B) Pelvic rotation during gait. (C) Pelvic tilt during gait. (D) Hip flexion, knee flexion and ankle dorsiflexion during gait.

the hip joint of the stance leg to in front of the hip joint of the stance leg during the swing. These movements are likely to increase the horizontal distance between the positions of the centre of gravity at toe-off and heel-strike of the swing leg without affecting the vertical excursion of the centre of gravity; this is likely to flatten the path of the centre of gravity in the median plane and reduce the change in direction of the centre of gravity required between steps.

2. Pelvic tilt (list, obliquity) during single support (see **Figure 7.26C**): relative hip adduction of the stance leg. This movement is likely to lower the centre of gravity and may decrease the vertical excursion of the centre of gravity.

3. Knee flexion during single stance (see **Figure 7.26D**): in association with hip flexion and ankle dorsiflexion, knee flexion during single stance is likely to lower the centre of gravity and may decrease the vertical excursion of the centre of gravity.

4. and 5. Rotation of the lead foot about the heel following heel strike, and extension of the metatarsophalangeal joints of the trail foot (in association will heel lift) during the latter half of the single support and the whole of double support (see **Figure 7.20F**): determinants 4 and 5 overlap during double support and, as shown by Hansen (2004), move the centre of pressure forward in a manner similar to that of a rolling wheel with radius approximately 30% of leg length (see **Figure 7.20F**). Like pelvic rotation, these movements are likely to reduce the change in direction of the centre of gravity required between steps.

6. Lateral displacement of the pelvis: maintaining a step width consistent with upright standing (lower legs vertical in the frontal plane) is likely to minimize mediolateral excursion of the centre of gravity.

Saunders et al (1953) did not present any empirical evidence in support of their six determinants. The purported effects of the six determinants, when considered independently, seem fairly logical and there is no doubt that the joint movements described in the determinants are present to a greater or lesser extent in normal gait (Orendurff et al 2004). The six determinants have been endorsed as important features in clinical gait analysis (Inman et al 1981; Perry 1992; Esquenazi & Talaty 2000). However, the six determinants do not occur independently of each other during gait and the effect of a particular determinant will depend upon the timings and ranges of motion of all of the determinants. There is empirical support for determinants 1, 4 and 5 (Hansen et al 2004; Adamczyk et al 2006) and determinant 6 (Donelan et al 2001). However, determinants 2 and 3 appear to have little or no effect on the vertical excursion of the centre of gravity and seem to be related to shock absorption following heel-strike (Gard & Childress 2001).

As the determinants refer to movements at joints, they are, when present in gait, kinematic features of the gait pattern, but not determinants of the pattern. The actual determinants of the gait pattern are the muscle forces that determine the movements of the joints and, in association with body weight, determine the ground reaction forces. The resultant of the ground reaction forces and body weight determine the force acting on the centre of gravity and, consequently, the velocity and displacement of the centre of gravity.

The inverted pendulum theory of gait

Figure 7.27A shows the movement of a pendulum. If the pendulum is stopped from swinging, it will hang vertically. If this position is regarded as the reference position for the measurement of gravitational potential energy ($h = 0$) and the whole mass of the pendulum is considered to be contained in the point mass at the end of the pendulum, then the pendulum will possess no gravitational potential energy or kinetic energy in this position. If the pendulum is rotated through an angle θ with respect to the vertical and held at rest, the work done on the pendulum will be equivalent to the gain in gravitational potential energy, i.e. $m.g.h$. If the pendulum is then released, it will oscillate for some considerable time and display a continuous transformation of mechanical energy from gravitation potential energy to kinetic energy in the downswings and from kinetic energy to gravitational potential energy in the upswings (see Figure 7.27B). In the absence of friction around its axis of rotation, and air resistance, the pendulum would oscillate forever, the total mechanical energy would be conserved, i.e. remain the same, and no energy expenditure would be needed to keep the pendulum oscillating.

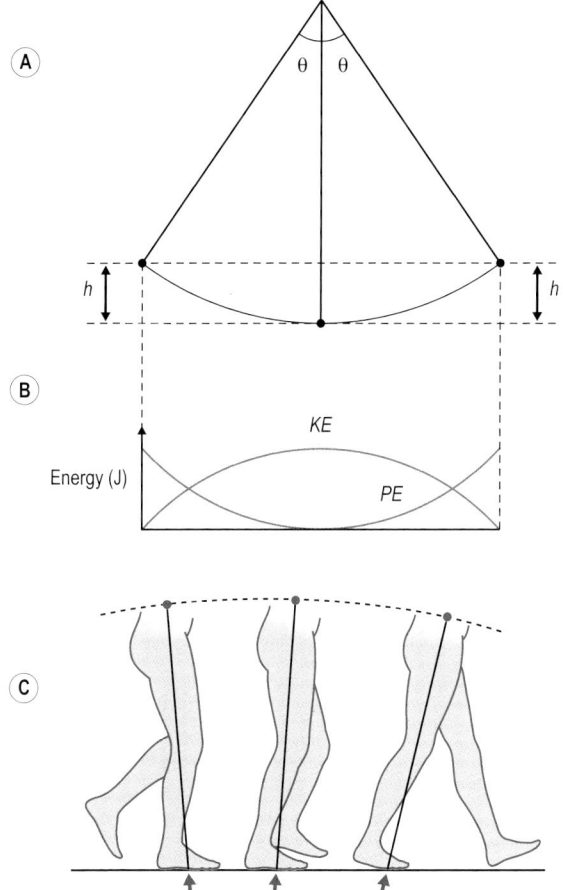

Figure 7.27 The movement of a pendulum. (A) In the absence of friction around the axis of rotation and air resistance, the total mechanical energy of the pendulum would be conserved. (B) Transformation of gravitational potential energy PE and kinetic energy KE of the pendulum during each swing. (C) Inverted pendulum movement of the body during the single support phase of gait.

In each single support phase of gait, the movement of the centre of gravity over the grounded foot is similar to that of an inverted pendulum (Cavagna & Margaria 1966) (see Figure 7.27C). Following toe off, the support leg remains fairly straight and behaves much like a strut, allowing kinetic energy to be transformed into gravitational potential energy during the period between toe-off and the high point of the centre of gravity around the middle of single support, and then allowing gravitational potential energy to be transformed into kinetic energy during the period between the high point of the centre of gravity and heel-strike. Just as the movement of the body over the grounded foot is similar to that of an inverted pendulum, the forward swing of the swing leg between toe-off and heel-strike is similar to that of an ordinary pendulum (see Figure 7.27C). Some energy expenditure is needed during single stance to keep the stance leg extended and to flex the swing leg to allow the foot to clear the ground. In addition, some energy will be lost due to the friction in joints and dissipation as heat in muscles. However, the movement of the body during single stance requires relatively little energy and much of the mechanical energy of the body is conserved due to pendulum motion (Cavagna et al 1977).

This is reflected in the forward velocity-time and vertical displacement-time curves of the centre of gravity (see Figure 7.24). As the vertical and mediolateral components of velocity are very small relative to the forward component, the magnitude of the forward component is almost the same as the resultant velocity of the centre of gravity. Figure 7.24 shows that the centre of gravity is at its lowest when the forward speed of the centre of gravity is at its highest, i.e. close to the middle of double support, and the centre of gravity is at its highest when the forward speed of the centre of gravity is at its lowest, i.e. close to the middle of single support. Consequently, the gravitational potential energy of the centre of gravity is at its lowest when the kinetic energy of the centre of gravity is at its highest and the gravitational potential energy of the centre of gravity is at its highest when the kinetic energy of the centre of gravity is at its lowest.

Step-to-step transitions during gait

The movement of the body as an inverted pendulum is a very effective and efficient way of transporting the body during single support. However, as the centre of gravity follows an arc path during single support (see Figure 7.27C), the direction of the centre of gravity needs to be changed during double support from forward, downward and medial with respect to the trail foot just before heel-strike of the lead foot (Figure 7.28A), to forward, upward and medial with respect to the lead foot just

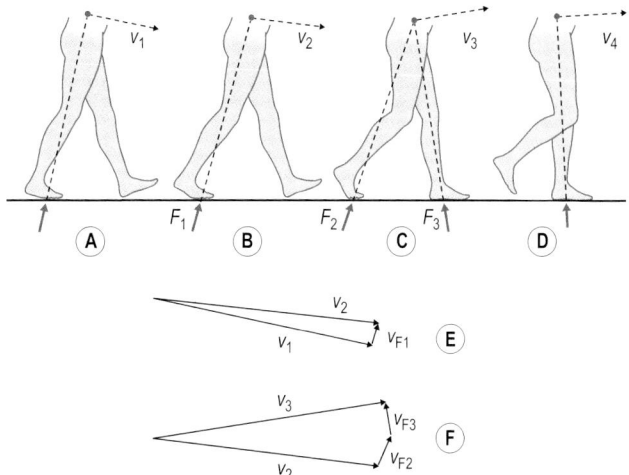

Figure 7.28 Step-to-step transition at a walking speed of 0.85 m/s. (A) Just before heel-strike. (B) Heel-strike. (C) Close to toe-off. (D) Just after toe-off. v_1, v_2, v_3 and v_4 are the velocities of the centre of gravity at A, B, C and D, respectively. F_1 and F_2, ground reaction forces on the right foot at B and C, respectively. F_3, ground reaction force on the left foot at C. v_{F1}, velocity vector due to the impulse of F_1. v_{F2}, velocity vector due to the impulse of F_2. v_{F3}, velocity vector due to the impulse of F_3.

after toe-off of the trail foot (see **Figure 7.28D**). The change in direction of the centre of gravity between steps is referred to by Kuo et al (2005) as the step-to-step transition. In the step-to-step transition, negative work has to be performed on the centre of gravity in order to arrest its downward velocity and medial velocity with respect to the trail foot. Negative work decreases the mechanical energy of the body. In order to maintain constant speed walking, the mechanical energy dissipated by the negative work must be replaced. This is achieved by positive work on the centre of gravity to generate upward velocity and medial velocity with respect to the lead foot.

During inverted pendulum motion in single support, the direction of the centre of gravity will be more-or-less perpendicular to the trail leg, i.e. perpendicular to the line joining the centre of pressure and the centre of gravity, as in **Figure 7.28A**. If the change in direction of the centre of gravity between steps is initiated too early, i.e. well before heel-strike, energy conservation due to pendulum motion will be decreased, resulting in an overall increase in energy expenditure. If the change in direction of the centre of gravity between steps is initiated too late, i.e. after

heel strike, the change in direction will be abrupt, similar to that in compass gait (see **Figure 7.25D**), which will also result in an overall increase in energy expenditure (Kuo 2007). Consequently, the step-to-step transition is normally initiated just before heel-strike (Donelan et al 2002). **Figure 7.28A and E** show the velocity vector v_1 of the centre of gravity just before heel-strike. In the period between the positions shown in **Figure 7.28A and B**, ankle plantar flexion of the trail leg initiates an upward–forward thrust along the line of the leg. This is the start of the push-off of the trail leg that terminates at toe-off. The impulse of the thrust generates a component of velocity v_{F1} of the centre of gravity in this direction (see **Figure 7.28E**). v_1, v_{F1} and their resultant v_2 are shown as vectors in **Figure 7.28E**. It is clear that v_{F1} changes the magnitude (very slightly greater) and direction (less downward) of the velocity of the centre of gravity during the first stage of the step-to-step transition (see **Figure 7.28B and E**). This stage terminates with heel-strike and is immediately followed by simultaneous negative work, largely by the lead leg, and positive work, largely by the trail leg. The negative work of the lead leg is equivalent to a change in velocity of the centre of gravity represented by the vector v_{F3} in **Figure 7.28F** and the positive work of the trail leg is equivalent to a change in velocity of the centre of gravity represented by the vector v_{F2} in **Figure 7.28F**. The resultant of v_2, v_{F2} and v_{F3} is v_3, the velocity of the centre of gravity just before toe-off.

Key Concepts

The sinusoidal movement of the centre of gravity in normal gait is a feature of an overall movement pattern that is optimal in terms of the energy expenditure needed to transport the body at a particular speed

Mechanical power of the legs during gait

In the SI system, the unit of work is the joule (J), after James Joule (1818–1889). One joule is the work done by a force of 1 newton (N) when it moves its point of application a distance of 1 metre (m) in the direction of the force, i.e. $1 J = 1 N.m$ (newton metre). In a concentric muscle contraction, if the muscle shortens a distance of 0.02 m (2 cm) while exerting an average force of 50 N, it will perform 1 J of positive work ($50 N \times 0.02 m = 1 J$). Similarly, in an eccentric muscle contraction, if the muscle lengthens a distance of 0.02 m while exerting an average force of 50 N, the muscle will perform 1 J of negative work, i.e. the muscle–tendon unit will absorb 1 J of strain energy.

Power is the rate of transformation of energy from one form to another. Mechanical power is the rate at which energy is transformed in the form of work. In the SI system, the unit of power is the watt (W), after James Watt (1736–1819). One watt is a work rate of 1 J/s. In a concentric muscle contraction, if the muscle performs 1 J of positive work in 10 ms (0.010 s), the average power output of the muscle during the contraction will be 100 W (= 1 J/0.01 s). Similarly, in an eccentric muscle contraction, if the muscle performs 1 J of negative work in 10 ms, the average rate at which the muscle absorbs strain energy during the contraction (sometimes referred to as power input) will be 100 W.

Figure 7.29A shows the power-time curves of the trail and lead legs during the gait cycle of a person walking at 1.25 m/s (Donelan et al 2002). The power-time curves show the rate at which the trail and lead legs do work on the centre of gravity. The areas between the curves and the time axis represent the work done (watts × seconds = joules) on the centre of gravity by the trail and lead legs in each stage of the gait cycle. Negative work (areas of the curves below the time axis) indicates a decrease in the mechanical energy of the body. Positive work (areas of the curves above the time axis) indicates an increase in the mechanical energy of the body. It is clear that most of the energy expenditure during gait (sum of the positive and negative work) occurs during the step-to-step transitions and that most of this work (60–70% of total energy expenditure) is done during double support (Donelan et al 2002). Approximately 94% of the negative work in double support is performed by the lead leg and approximately 97% of the positive work in double support is performed by the trail leg (Donelan et al 2002).

Figure 7.29B shows the net power-time curve for both legs combined. The net power-time curve shows that there are five phases of alternating positive and negative power during each step. Phase 1 is a period of net positive power, which takes up the first half of double support (Figure 7.30A and B). Positive work is performed by the trail leg in a continuation of the push-off that was initiated before heel-strike of the lead leg. Positive work is in the form of active positive work, i.e. concentric muscle contractions and, possibly, passive positive work in the form of release of strain energy that was stored in muscle–tendon units (especially the ankle plantar flexors) prior to the start of push-off. Negative work is performed by the lead leg in the form of active negative work, i.e. eccentric muscle contractions of leg extensors and intrinsic muscles of the foot, and passive negative work as a result of deformation of passive structures including the shoe, the heel pad and the arches of the feet. Active and passive negative work will result, at least momentarily, in storage of strain energy. The net effect of the work done on the centre of gravity during this phase

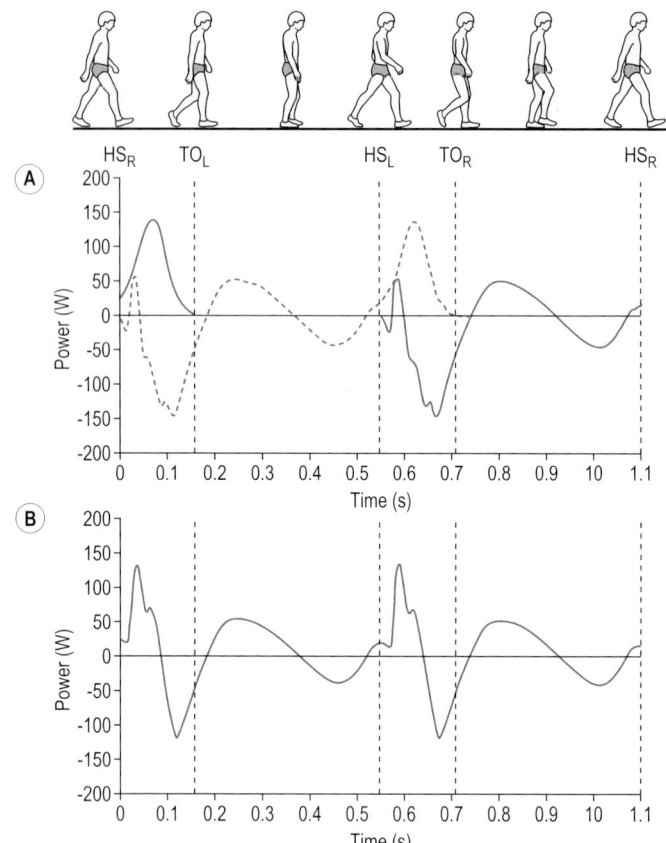

Figure 7.29 (A) Power-time curves of the work done by the left and right legs on the centre of gravity by a healthy adult walking at 1.25 m/s during one complete gait cycle (heel-strike to heel-strike of the right foot). (B) Power-time curve of the work done by the legs (both legs combined) for the same cycle in A. HS$_R$, heel-strike right; HR$_L$, heel-strike left; TO$_R$, toe-off right; TO$_L$, toe-off left. Based on data from Donelan JM, Kram R, Kuo AD (2002) Simultaneous positive and negative external work in human walking. Journal of Biomechanics 35:117–24.

is an increase in forward velocity, a decrease in the downward velocity to zero, a slight increase in medial velocity with respect to the trail foot, and a decrease in gravitational potential energy (see **Figure 7.24**).

Phase 2 is a period of net negative work from the middle of double support to just after toe-off of the trail foot (see **Figure 7.30B to D**). Positive work

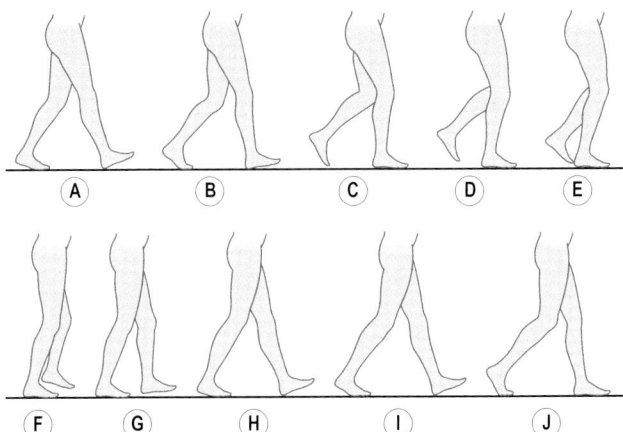

Figure 7.30 Movement of the legs during the period from heel-strike of the right foot to just after heel-strike of the left foot of a healthy adult walking at 0.85 m/s.

is performed by the trail leg in the final phase of the push-off. Negative work is performed by the lead leg and involves slight knee flexion and ankle dorsiflexion. The net effect of the work done on the centre of gravity during this phase is a decrease in forward velocity, generation of upward velocity, a slight decrease in medial velocity with respect to the trail foot and an increase in gravitational potential energy (see **Figure 7.24**).

Phases 1 and 2, dominated by the step-to-step transition, are referred to collectively by Kuo et al (2005) as the collision phase.

Phase 3 is a period of positive work by the stance leg from just after toe-off of the trail foot to midstance (when the stance leg is approximately vertical), and corresponds to the upward arc of the inverted pendulum (see **Figure 7.30D to F**). Positive work is reflected in slight knee extension. The effect of the positive work done on the centre of gravity during this phase is a continued decrease in forward velocity, a slight increase followed by decrease in the upward vertical velocity to zero, decrease in medial velocity with respect to the trail (swing) foot to zero and an increase in gravitational potential energy (see **Figure 7.24**). This phase is referred to by Kuo et al (2005) as the rebound phase.

Phase 4 is a period of negative work by the stance leg from midstance to the start of the push-off and corresponds to the downward arc of the inverted pendulum (see **Figure 7.30F to H**). During this period, strain energy is stored in the ankle plantar flexors. The effect of the negative work done on the centre of gravity during this phase is an increase in forward velocity, generation of downward vertical velocity, generation of medial velocity with respect to the lead foot and a decrease in gravitational potential energy (see **Figure 7.24**). This phase is referred to by Kuo et al (2005) as the preload phase.

Phase 5 is a very brief period of positive work by the stance leg from the start of the push-off to heel-strike (see **Figure 7.30H to I**). It is the start of the step-to-step transition and is probably associated with the release of strain energy in the ankle plantar flexors. Kuo et al (2005) refers to the period of continuous positive work by the trial leg during phases 5, 1 and 2 as the push-off.

Theories of foot function during gait

Historically, there are three main theories of foot function upon which podiatric assessment and treatment of skin and musculoskeletal disorders of the foot have been based: subtalar joint neutral theory (Root et al 1977), sagittal plane facilitation theory (Dananberg 1986) and tissue stress theory (McPoil & Hunt 1995). In all three approaches, treatment is based on the production of a customized foot orthosis, i.e. a shoe insert designed to restore normal movement of the foot during gait and, in doing so, alleviate pain/discomfort in the foot and/or other parts of the body such as the knee, hip and lower back (Sobel et al 1999). Orthotics is the field of study concerned with the design and manufacture of orthoses. An orthosis is a device that is applied to a part of the body to correct deformity, improve function or relieve symptoms of a disease by supporting or assisting the neuromusculoskeletal system (Wikipedia 2007). The term is derived from the Greek word *orthos* meaning straight or correct.

The subtalar joint neutral approach is based on the belief that: (1) a foot functions normally during gait when the subtalar joint is in neutral just following heel-strike and at midstance; (2) in a normal foot, weightbearing and non-weightbearing, the plane of the plantar surface of the forefoot is perpendicular to the vertical bisection of the posterior aspect of the calcaneus when the subtalar joint is in neutral; and (3) in a normal foot, the calcaneal inclination angle in the frontal plane is zero when the subtalar joint is in neutral, i.e. the vertical bisection of the posterior aspect of the calcaneus is in line with the long axis of the lower leg.

In the subtalar joint neutral approach, the foot is assessed by casting the foot in the subtalar joint neutral position while non-weightbearing in order to capture the forefoot–rearfoot orientation and the calcaneal inclination angle. This information is then used to produce an orthosis consisting of a rigid base with forefoot and/or rearfoot posting. There would appear to be no empirical evidence from well-controlled studies in support of the theory or method of treatment, and some research indicates that the theory is not valid (McPoil & Cornwall 1996; Pierrynowski & Smith 1996). However, anecdotal reports over many years in support of the method of treatment have resulted in the subtalar joint neutral approach becoming well established (Harradine et al 2003).

The sagittal plane facilitation approach is based on the belief that each foot functions as a pivot in the sagittal plane that facilitates energy conservation by smoothing the path of the centre of gravity. As described earlier, it has been shown that in normal gait, the forward movement of the centre of pressure on the foot is similar to that of a rolling wheel with radius approximately 30% of leg length (see **Figure 7.20F**) (Hansen et al 2004; Adamczyk et al 2006). Furthermore, it has been shown that in normal gait, the timing of the leg actions during step-to-step transitions tends to minimize energy expenditure (see **Figures 7.28 and 7.29**) (Donelan et al 2002; Kuo et al 2005). Consequently, there is empirical support for the sagittal plane facilitation theory. Whereas the forefoot–rearfoot orientation and calcaneal inclination angle are central to the subtalar joint neutral theory, rotation of the foot about three consecutive sagittal plane 'rockers', the heel (heel-strike to foot flat), the ankle joint (dorsiflexion during the inverted pendulum phase) and the metatarsophalangeal joints (extension from heel-off to just before toe-off) is central to the sagittal plane facilitation theory (**Figure 7.31**). In the sagittal plane facilitation theory, smooth movement during each rocker and smooth transitions between the three rockers is regarded as essential to normal foot function. Restricted ranges of motion in the ankle and metatarsophalangeal joints is said to result in

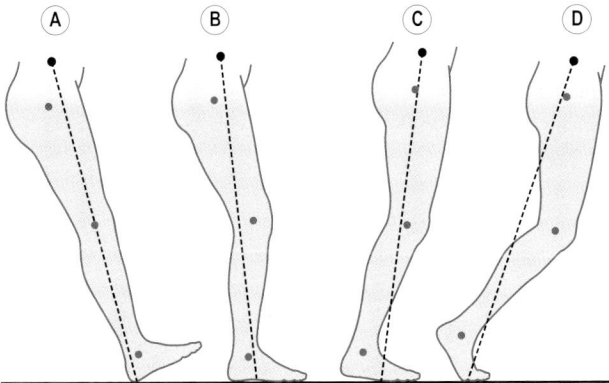

Figure 7.31 The three rockers in the sagittal plane facilitation theory of foot function during gait. (A–B) Rocker 1. (B–C) Rocker 2. (C–D) Rocker 3.

'sagittal plane blockade' and, consequently, compensatory movements and increased energy expenditure.

In the sagittal plane facilitation approach, the movement of the ankle and foot is assessed qualitatively (observation of slow-motion video of gait) and quantitatively (foot and ankle ranges of motion, muscle strength, leg-length discrepancy, plantar pressure during gait) (Dananberg & Guiliano 1999). The results of the assessment are then used to design and subsequently refine a semi-rigid orthosis. Unfortunately, no detailed descriptions of the methods of assessment or stages in the design of the orthosis have been published. Consequently, whereas there is empirical support for the theory underlying the sagittal plane facilitation approach, there is very limited evidence of the effectiveness of the treatment (Harradine et al 2003).

The tissue stress approach is based on the belief that: (1) abnormal foot shape, especially flat foot (pes planus) displaces the axis of the subtalar joint (orientation of the axis and position of the axis with respect to the centre of pressure) from its normal orientation/position; (2) the displacement of the subtalar joint axis results in the ground reaction force exerting an excessive moment about the subtalar joint; and (3) the excessive ground reaction force moment results in excessive loading on the structures of the foot (muscles, tendons, ligaments, joint surfaces) that counteract the ground reaction force moment.

There is no doubt that musculoskeletal components are damaged by excessive loading, but a change in the position of the centre of pressure that accompanies a change in the shape of the foot during stance will alter the moment of the ground reaction force and, consequently, the musculoskeletal response, about all of the joints in the foot, not just the subtalar joint. At present there would appear to be no reliable method of determining, simultaneously, the orientation/position of the subtalar joint and centre of pressure during gait and, therefore, no way of accurately assessing the effect of an orthosis on the moment of the ground reaction force about the subtalar joint and the associated musculoskeletal response.

Assessment in the tissue stress approach would appear to be largely qualitative (based on observation) with the main focus of attention on the extent and duration of pronation during gait. This information is then used to produce an orthosis consisting of rigid or semi-rigid base with forefoot and/or rearfoot posting. As with the sagittal plane facilitation approach, no detailed descriptions of the methods of assessment or stages in the design of the orthosis in the tissue stress approach have been published. Consequently, there is no reported evidence from controlled studies of the effectiveness of the treatment (Harradine et al 2003).

Key Concepts

There are three main theories of foot function underlying podiatric clinical practice: subtalar joint neutral theory, sagittal plane facilitation theory and tissue stress theory. In all three approaches, treatment is based on the production of a customized foot orthosis

Effects of orthoses on foot function in gait

As the subtalar joint neutral approach, sagittal plane facilitation approach and tissue stress approach are all prominent in podiatric clinical practice, it is reasonable to assume that they are all effective in relieving pain and discomfort for the majority of patients. Consequently, it is also reasonable to assume that the methods of treatment, i.e. the types of customised orthoses that result from the three approaches, have common characteristics that improve foot function and, consequently, remove or reduce abnormal loading on particular components of the foot.

The main functions of the foot in gait are to: (1) provide shock absorption during the impact phase following heel-strike; (2) provide a stable base of support; (3) facilitate energy conservation; and (4) provide propulsion during push-off. Functions (1)–(3) overlap in the first half of the stance phase and functions (2)–(4) overlap during the second half of the stance phase. Semi-rigid orthoses, customized and non-customized, are frequently prescribed to assist with shock absorption in walking and running, and orthoses that support the medial longitudinal arch of the foot, a feature of most orthoses, are likely to improve foot function in relation to the provision of a stable base of support, energy conservation and propulsion (Kogler et al 1996; Nigg 2001).

Shock absorption following heel-strike

During the impact phase following heel-strike (see Figure 7.30A and B), the centre of gravity experiences a decrease in mechanical energy (net decrease in gravitational potential energy and kinetic energy). The prime function of the musculoskeletal system as a whole, and the foot in particular, during the impact phase is shock absorption, i.e. dissipation of the mechanical energy lost by the centre of gravity in a way that allows the foot to establish good contact with the floor and not tend to bounce away from the floor as would a ball. This is normally achieved by compression of the heel of the shoe, compression of the insole and/or orthosis inside the shoe and compression of the heel pad, i.e. the thick layer of fat in the skin beneath the calcaneus. The impact phase is followed by a loading phase (see **Figure 7.30B–C**) in which the whole of body weight is transmitted to the floor. In the loading phase, eccentric action of the ankle dorsiflexors contributes to shock absorption by lowering the plantar surface of the foot to the floor under control. The shock absorption phase (impact and loading phases) corresponds to the first rocker in the sagittal plane facilitation theory of foot function, i.e. from heel-strike to foot flat (see **Figure 7.31A–B**).

There is a layer of fat, referred to as the subcutaneous fat layer, beneath the skin all over the body. About half of all the fat in the adult body is located in the subcutaneous fat layer such that total body fat can be reliably estimated from the sum of skinfold measurements taken from different parts of the body (Lohman 1981). The thickness of the subcutaneous fat layer varies throughout the body and is particularly thick over the inferior aspects of the calcaneus (about 18 mm in an adult) and heads of the metatarsals where it protects the bones against impact with the ground following heel-strike (Bojsen-Møller 1999). The subcutaneous fat layer over the inferior aspect of the calcaneus is usually referred to as the heel pad (**Figure 7.32**).

The blood vessels in the heel pad normally form a large number of plexuses (networks) that allow blood to flow freely in many directions in response to loading. In response to impact at heel-strike, compression of the heel pad results in compression of the blood vessels in the heel, which forces blood away from the impact site. Consequently, much of the mechanical energy of the centre of gravity that is lost during the

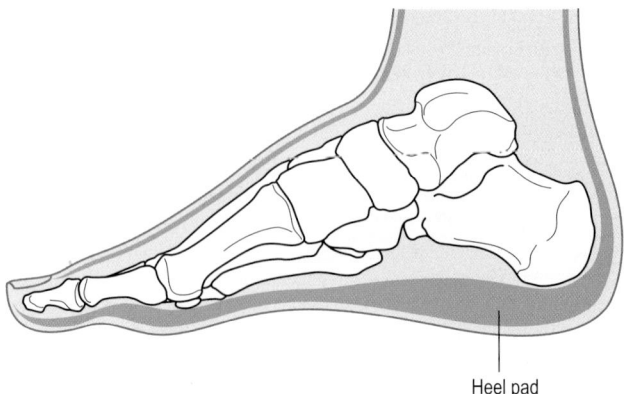

Heel pad

Figure 7.32 The medial aspect of the right foot showing the subcutaneous layer of fat and heel pad.

impact phase is transformed as increased kinetic energy of blood flowing away from the heel, i.e. compression of the heel pad acts as a hydraulic pump that assists the return of venous blood to the heart. Restoration of the thickness of the heel pad following impact is due largely to blood flowing back into the heel pad rather than the release of strain energy in the heel pad. Consequently, in response to impact loading, the heel pad acts as a shock absorption system (low resilience energy absorption system) that prevents the foot from bouncing away from the floor.

The windlass mechanism of the foot

As described earlier, the plantar aponeurosis spans the whole of the tarsus and metatarsus by linking the inferior aspect of the calcaneus with the plantar surfaces of the bases of the proximal phalanges of the toes (see Figures 7.10 and 7.11A and D). Consequently, extension of the metatarsophalangeal joints winds the plantar aponeurosis around the heads of the metatarsals, like a cable being wound around a windlass, which simultaneously raises the longitudinal arch (Figure 7.33). This action is referred to as the windlass mechanism of the foot (Hicks 1954). Flexion of the metatarsophalangeal joints unwinds the plantar aponeurosis and lowers the longitudinal arch; this action is referred to as the reverse windlass (Aquino & Payne 2000).

Reverse windlass action is a feature of the loading phase and much of the single support phase (up to heel-off) (see Figure 7.30C–G). During this period, the rearfoot complex normally pronates, which unwinds the plantar aponeurosis and lowers the longitudinal arch. In the foot flat position, the tension in the plantar aponeurosis exerts a flexor moment on the proximal phalanges (pushes the pads of the toes against the ground), which extends the length of the base of support and, consequently, reduces the pressure on the plantar surfaces of the heads of the metatarsals. In addition, the tension in the plantar aponeurosis, in association with tension in the intrinsic muscles, prevents excessive flattening of the longitudinal and transverse arches and provides a stable base of support. The period from foot-flat to heel-off corresponds to the second rocker in the sagittal plane facilitation theory of foot function (see Figure 7.31B–C).

Windlass action is a feature of the push-off, i.e. from heel-off to just before toe-off (see Figure 7.30G–J). During this period, the rearfoot complex normally supinates in association with extension of the metatarsophalangeal joints. These actions raise the longitudinal arch, which stabilizes the foot and provides a firm base of support for the push-off. The period from heel-off to just before toe-off corresponds to the third rocker in the sagittal plane facilitation theory of foot function (see Figure 7.31C–D).

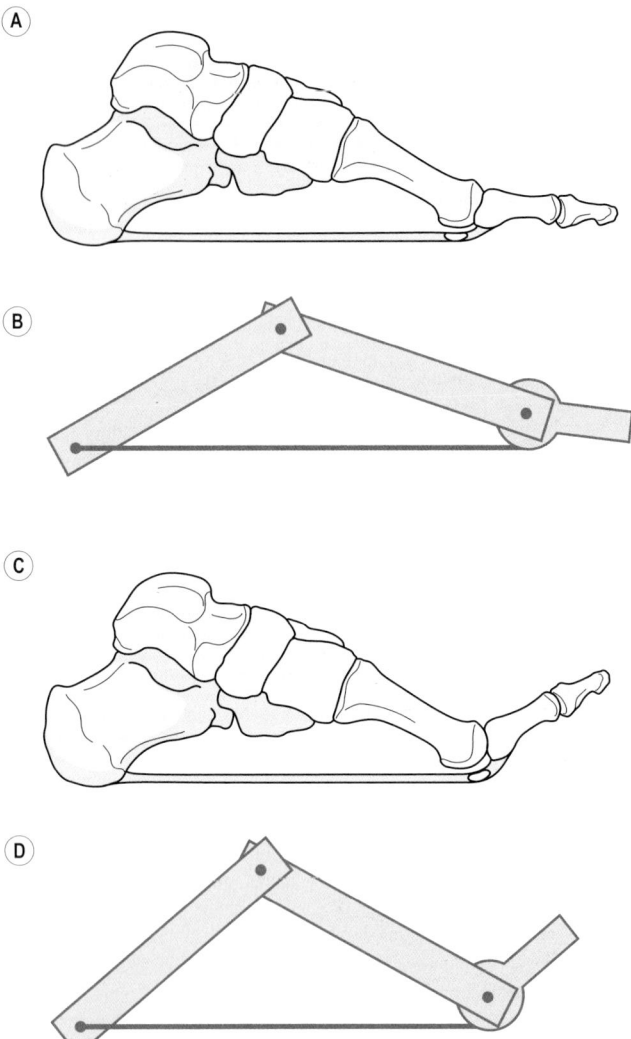

Figure 7.33 The windlass mechanism of the foot. (A) Reverse windlass orientation of the foot. (B) Model of a windlass in the unwound position. (C) Windlass orientation of the foot. (D) Model of a windlass in the wound position.

The effectiveness of the windlass mechanism, i.e. the speed of initiation of the arch raise and the extent to which the arch is raised, depends upon the moment arm of the plantar aponeurosis about the metatarsophalangeal joints (Bojsen-Møller 1979). For a given extension of the metatarsophalangeal joints, the greater the moment arm of the plantar aponeurosis, the more effective the windlass. In an adult, the radius of the head of the first metatarsal (in the sagittal plane) is approximately 10 mm and the radius of the heads of the second to fifth metatarsals is approximately 8 mm. Not only is the radius of the first metatarsal larger than that of the other metatarsals, but each of the slips of the plantar aponeurosis to the proximal phalanx of the hallux contains a sesamoid bone that forms a synovial joint with the head of the first metatarsal (Figure 7.34). The sesamoid bones increase the moment arm of the plantar aponeurosis about the head of the first metatarsal to about 15 mm. Consequently, the plantar aponeurosis is a much more effective windlass about the first metatarsal head than about the other metatarsal heads.

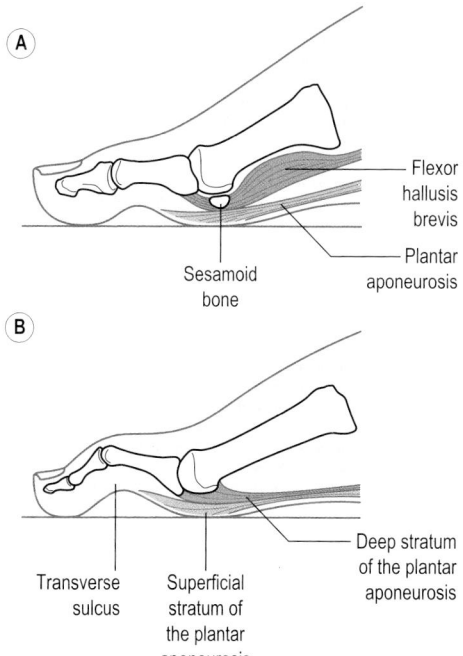

Figure 7.34 Sagittal sections through the first (A) and second (B) metatarsophalangeal joints.

Bojsen-Møller (1979) pointed out that the head of the second meta-tarsal is normally slightly anterior to the heads of the first and third meta-tarsals such that the metatarsophalangeal joints do not share a common axis of extension/flexion. Bojsen-Møller (1979) referred to the axis of extension/flexion through the first and second metatarsophalangeal joints as the transverse axis, and the axis of extension/flexion through the sec-ond to fifth metatarsophalangeal joints as the oblique axis (Figure 7.35A). Bojsen-Møller (1979) demonstrated that during the push-off phase in gait, metatarsophalangeal joint extension normally occurs primarily about the transverse axis, such that the foot tends to move in a sagittal plane and the centre of pressure follows a path from the heel to the medial part

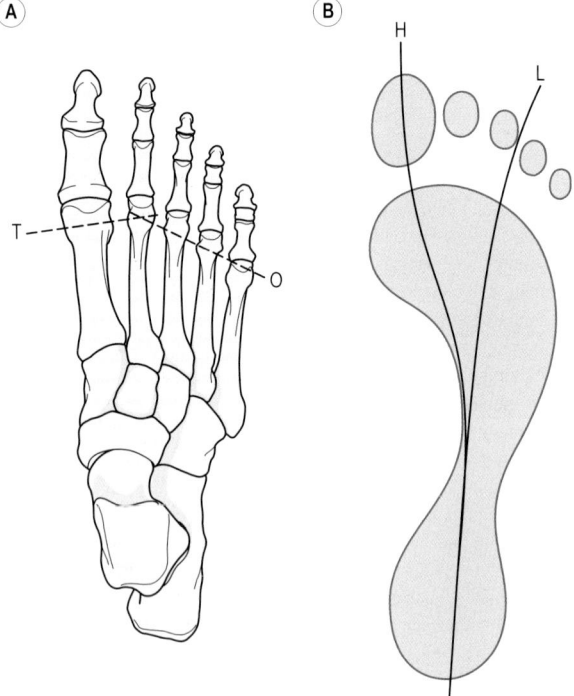

Figure 7.35 (A) Transverse axis of flexion/extension of the first and second metatarsophalan-geal joints **T**, and oblique axis of flexion/extension of the second to fifth metatarsophalangeal joints **O**. (B) Path of the centre of pressure during high gear **H** and low gear **L** windlass actions.

of the ball of the foot and, finally, to the hallux (see Figure 7.35B). In contrast, push-off over the oblique axis is associated with forefoot adduction in relation to the rearfoot, and a centre of pressure path that passes from the heel to the lateral part of the ball of the foot and, finally, to the third and fourth toes (see Figure 7.35B). Owing to the greater moment arm of the plantar aponeurosis about the first metatarsophalangeal joint, Bojsen-Møller (1979) described extension of the metatarsophalangeal joints about the transverse axis as high gear windlass action and extension of the metatarsophalangeal joints about the oblique axis as low gear windlass action. High gear windlass action will stabilize the foot more effectively than low gear windlass action during push-off and, consequently, reduce the risk of excessive overload to particular plantar ligaments, intrinsic muscle–tendon units, intertarsal joints and tarsometatarsal joints that may arise from prolonged pronation.

The common feature of orthoses shown to be effective in the treatment of disorders associated with excessive and/or prolonged pronation may be support for the medial longitudinal arch of the foot, especially support for the apex of the medial longitudinal arch, i.e. the region of the sustentaculum tali, inferior aspect of the head of the talus, and inferior aspect of the navicular. The apex of the medial longitudinal arch is medial to the medial aspect of the plantar aponeurosis. Orthoses that support the apex of the medial longitudinal arch, but do not encroach on the medial aspect of the plantar aponeurosis, have been shown to reduce the strain on the plantar aponeurosis during standing (Kogler et al 1996). In contrast, orthoses that do not provide much support for the apex of the medial longitudinal arch, but do encroach on the medial aspect of the plantar aponeurosis, are likely to increase the strain on the plantar aponeurosis (like pulling on a bowstring). A semi-rigid orthosis that reduces the strain on the plantar aponeurosis is likely to improve the effectiveness of the windlass.

Hallux valgus

The metatarsophalangeal joints are biaxial condyloid joints that allow flexion–extension and abduction–adduction; the convex condylar surface of the head of each metatarsal articulates with the concave condylar surface of the base of the corresponding proximal phalanx (Figure 7.36A). The medial and lateral aspects of the plantar surface of the head of the first metatarsal normally form shallow parallel grooves either side of a central sagittal ridge called the crista (see Figure 7.36B). The medial and lateral sesamoid bones articulate with the medial and lateral grooves, respectively. The sesamoid bones slide forward in their respective grooves

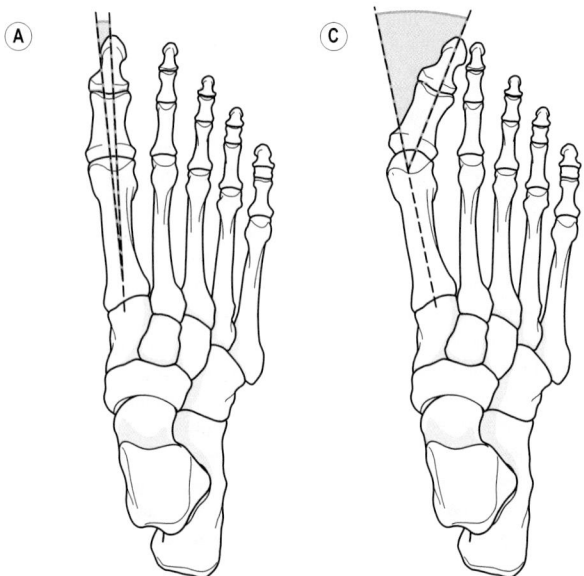

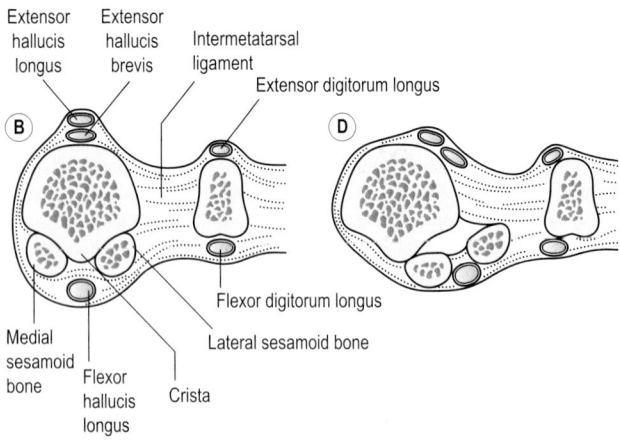

Figure 7.36 Hallux valgus. (A) Superior aspect of the right foot with a hallux abductus angle of approximately 5°. (B) Posterior aspect of a coronal section through the head of the first metatarsal corresponding to A. (C) Superior aspect of the right foot with a hallux abductus angle of approximately 35°. (D) Posterior aspect of a coronal section through the head of the first metatarsal corresponding to C.

during extension (windlass action) and backward during flexion (reverse windlass action) of the first metatarsophalangeal joint.

Windlass action about the first metatarsophalangeal joint is most effective when: (1) the sesamoids are located within their respective grooves, which maximizes the moment arm of the windlass; and (2) the long axes of the first metatarsal and phalanges of the hallux are in line, such that the axis of flexion/extension of the first metatarsophalangeal joint is perpendicular to the long axes of the bones (see **Figure 7.36A and B**). Not surprisingly, this would appear to be the normal orientation of the joint, as significant non-alignment of the first metatarsal and proximal phalanx seems to be rare in children (Kilmartin et al 1991).

With increase in age, many people develop hallux valgus, also referred to as hallux abducto valgus (Thomas & Barrington 2003). Hallux valgus is a complex progressive condition that is characterized by lateral deviation (valgus, abduction) of the hallux and medial deviation of the first metatarsophalangeal joint. Unless treated, hallux valgus results in a progressive increase in the hallux abductus angle, i.e. the angle between the long axes of the first metatarsal and proximal phalanx in the transverse plane (see **Figure 7.36C**). When the hallux abductus angle is less than 15°, the condition tends to be asymptomatic. However, increases in the hallux abductus angle above 15° tend to be associated with increasing pain and discomfort around the first metatarsophalangeal joint (Menz & Lord 2005; Easley & Trnka 2007). Whereas the cause of hallux valgus may be congential in some cases, there is general agreement that the main cause is prolonged use of ill-fitting footwear, especially shoes with three particular characteristics: pointed toe regions, insufficient length and high heels (Mays 2005).

Relative to the first metatarsophalangeal joint, any increase in the hallux abductus angle will tend to displace laterally the lines of action of the plantar aponeurosis and tendons of the intrinsic and extrinsic muscles that cross over the first metatarsophalangeal joint from the metatarsal to the hallux. Consequently, the sesamoid bones will also be displaced laterally relative to the first metatarsophalangeal joint, resulting in subluxation of the joints between the sesamoid bones and the head of the first metatarsal (see **Figure 7.36D**). Subluxation of these joints is likely to increase the pressure between the medial sesamoid and the crista of the first metatarsal head, and decrease the pressure between the lateral sesamoid and the first metatarsal head. Subluxation of these joints will also decrease the mechanical efficiency of the windlass about the first metatarsophalangeal joint and, consequently, tend to result in increased force in the muscles supporting the arches during weightbearing in order to compensate for the loss in mechanical efficiency of the windlass. The increased muscle

force will tend to increase the hallux abductus angle and, consequently, increase: (1) the pressure on the articular surfaces between the medial sesamoid and the crista; (2) the strain on the medial ligaments and intersesamoid ligament of the first metatarsophalangeal joint; (3) the strain on the intertransverse ligament and metatarsophalangeal ligament between the first and second metatarsophalangeal joints; and (4) the pressure exerted by the shoe on the medial aspect of the first metatarsophalangeal joint (Tanaka et al 1997). If this abnormal pattern of loading is prolonged, the hallux valgus will become progressively worse and result in permanent deformity due to lengthening of the medial ligaments of the first metatarsophalangeal joint and shortening (contractures) of the laterally displaced plantar aponeurosis and related muscles.

Prolonged excessive pressure between the medial sesamoid and the crista is likely to result in progressive erosion of both surfaces and/or osteoarthritis. Prolonged excessive pressure on the medial aspect of the first metatarsal joint is likely to result in the development of a bursa, commonly known as a bunion. The painful conditions associated with hallux valgus, and permanent loss of high-gear windlass action due to hallux valgus, is likely to result in inefficient gait, impaired balance and an increased risk of falling, especially in the elderly (Menz & Lord 2005).

Key Concepts

The aetiology of hallux valgus is unknown. Whilst mechanical dysfunction of the foot is perceived to be implicated by some clinicans, this remains unproven. However, there is general agreement that the main cause of hallux valgus is prolonged use of ill-fitting footwear, especially shoes with three particular characteristics: pointed toe regions, insufficient length and high heels

Review questions

1. Describe the rearfoot complex.
2. Differentiate between the following:
   - Supination and pronation of the foot
   - Passive and active arch support mechanisms
   - Intrinsic and extrinsic muscles of the foot
   - Direct arch raiser and indirect arch raiser
   - Energy and work
   - Positive work and negative work

- Gravitational potential energy, kinetic energy and strain energy
- Shock absorption and energy absorption
- Windlass and reverse windlass actions
- Transverse and oblique axes of windlass action
- High-gear and low-gear windlass actions

3. Define gait cycle, step, stride, podiatric biomechanics, compass gait and step-to-step transition.
4. Describe the six determinants of gait theory and the inverted pendulum theory of gait.
5. Describe the subtalar joint neutral theory, sagittal plane facilitation theory and tissue stress theory of foot function during gait.
6. Describe the main functions of the foot during gait.
7. Describe the effects of hallux valgus on foot function.

References

Adamczyk PG, Collins SH, Kuo AD (2006) The advantages of a rolling foot in human walking. Journal of Experimental Biology 209:3953–63.

Alexander RM (1995) Simple models of human movement. Applied Mechanics Reviews 48:461–69.

Aquino A, Payne C (2000) The role of the reverse windlass mechanism in foot pathology. Australasian Journal of Podiatric Medicine 34:32–4.

Åström M, Arvidson T (1995) Alignment and joint motion in the normal foot. Journal of Sports Physical Therapy 22:216–22.

Bertram JE, Ruina A (2001) Multiple walking speed-frequency relations are predicted by constrained optimisation. Journal of Theoretical Biology 209:445–53.

Bojsen-Møller F (1979) Calcaneocuboid joint and stability of the longitudinal arch of the foot at high and low gear push off. Journal of Anatomy 129:165–76.

Bojsen-Møller F (1999) Biomechanics of the heel pad and plantar aponeurosis. In: Ranawat CS, Positano RG (eds) Disorders of the heel, rearfoot, and ankle. Churchill Livingstone, Edinburgh, pp 137–43.

Boruta PM, Bishop JO, Braly G, Tullos HS (1990) Acute lateral ankle ligament injuries: a literature review. Fott and Ankle 11:107–13.

Bowden PD, Bowker P (1995) The alignment of the rearfoot complex axis as a factor in the development of running induced patellofemoral pain. Journal of British Podiatric Medicine 50:114–18.

Briggs PJ (2005) The structure and function of the foot in relation to injury. Current Orthopaedics 19:85–93.

Cavagna GA, Margaria R (1966) Mechanics of walking. Journal of Applied Physiology 21:271–78.

Cavagna GA, Heglund NC, Taylor CR (1977) Mechanical work in terrestrial locomotion: two basic mechanisms for minimizing energy expenditure. American Journal of Physiology 233:R243–R261.

Cheung JT, Zhang M, An K (2004) Effect of plantar fascia stiffness on the biomechanical responses of the ankle-foot complex. Clinical Biomechanics 19:839–46.

Crary JL, Hollis JM, Manoli A (2003) The effect of plantar fascia release on strain in the spring and long plantar ligaments. Foot and Ankle International 24:245–50.

Czerniecki JM (1988) Foot and ankle biomechanics in walking and running: a review. American Journal of Physical Medicine and Rehabilitation 67:246–52.

Dananberg HJ (1986) Functional hallux limitus and its relationship to gait efficiency. Journal of the American Podiatric Medical Association 76:648–52.

Dananberg HJ, Guiliano M (1999) Chronic low-back pain and its response to custom-made orthoses. Journal of the American Podiatric Medical Association 80:109–17.

Donelan JM, Kram R, Kuo AD (2001) Mechanical and metabolic determinants of the preferred step width in human walking. Proceedings of the Royal Society of London: Biological Sciences 268:1985–92.

Donelan JM, Kram R, Kuo AD (2002) Simultaneous positive and negative external work in human walking. Journal of Biomechanics 35:117–24.

Downing BS, Klein BS, D'Amico JS (1978) The axis of motion of the rearfoot complex. Journal of the American Podiatric Association 68:484–99.

Easley ME, Trnka H-J (2007) Current concepts review: hallux valgus part 1: pathomechanics, clinical assessment, and nonoperative management. Foot & Ankle International 28:654–9.

Esquenazi A, Talaty M (2000) Normal and pathological gait analysis. In: Grabois M, Garrison SJ, Hart KA, Lehmkuhl LD (eds) Physical medicine and rehabilitation: the complete approach. Blackwell Science, Malden, MD, pp 242–62.

Gage JR (1990) An overview of normal walking. In: Greene WB (ed) Instructional course lectures, vol 29: American Academy of Orthopedic Surgeons. CV Mosby, St Louis, pp 291–303.

Gard SA, Childress DS (2001) What determines the vertical displacement of the body during normal walking? Journal of Prosthetics and Orthotics 3:64–67.

Gregor RJ (1993) Skeletal muscle mechanics and movement. In: Grabiner MD (ed) Current issues in biomechanics. Human Kinetics, Champaign, IL, pp 171–211.

Hansen AD, Childress DS, Knox EH (2004) Roll-over shapes of human locomotor systems: effects of walking speed. Clinical Biomechanics 19:407–14.

Harradine PD, Bevan LS, Carter N (2003) Gait dysfunction and podiatric therapy – part 1: foot-based models and orthotic management. British Journal of Podiatry 6:5–11.

Harradine PD, Bevan LS, Carter N (2006) An overview of podiatric biomechanics theory and its relation to selected gait dysfunction. Physiotherapy 92:122–7.

Holt KG, Jeng S-F, Ratcliffe RJ, Hamill J (1995) Energetic cost and stability during human walking at the preferred stride frequency. Journal of Motor Behavior 27:164–78.

Hicks JH (1954) The mechanics of the foot. II. The plantar aponeurosis and the arch. Journal of Anatomy 88:25–31.

Hicks JH (1961) The three weightbearing mechanisms of the foot. In: Evans FG (ed) Biomechanical studies of the musculoskeletal system. Charles C. Thomas, Springfield, IL, pp 161–91.

Inman VT (1976) Joints of the ankle. Williams & Wilkins, Baltimore.

Inman VT, Ralston HJ, Todd F (1981) Human walking. Williams & Wilkins, Baltimore.

Jeng S-F, Liao H-F, Lai J-S, Hou J-W (1997) Optimization of walking in children. Medicine and Science in Sports and Exercise 29:370–6.

Kent M (2006) Oxford dictionary of sports science and medicine, 3rd edn. Oxford University Press, Oxford.

Kilmartin TE, Barrington RL, Wallace AW (1991) Metatarsus primus varus, a statistical study. Journal of Bone and Joint Surgery 73B:937–40.

Kitaoka HB, Ahn T-K, Luo ZP, An K-N (1997a) Stability of the arch of the foot. Foot Ankle International 18:644–8.

Kitaoka HB, Luo ZP, An K-N (1997b) Three-dimensional analysis of normal ankle and foot mobility. American Journal of Sports Medicine 25:238–42.

Kogler GF, Solomonidis SE, Paul JP (1996) Biomechanics of longitudinal arch support mechanisms in foot orthoses and their effect on plantar aponeurosis strain. Clinical Biomechanics 11:243–52.

Kuo AD (2007) The six determinants of gait and the inverted pendulum analogy: a dynamic walking perspective. Human Movement Science 26:617–56.

Kuo AD, Donelan JM, Ruina A (2005) Energetic consequences of walking like an inverted pendulum; step-to-step transitions. Exercise and Sport Sciences Reviews 33:88–97.

Lassiter T, Malone T, Garrick J (1989) Injury to the lateral ligaments of the ankle. Orthopedic Clinics of North America 20:629–40.

Lohman TG (1981) Skinfolds and body density and their relation to body fatness: a review. Human Biology 53:181–225.

Mays SA (2005) Paleopathological study of hallux valgus. American Journal of Physical Anthropology 126:139–49.

McPoil TG, Cornwall MW (1996) Relationship between three static angles of the rearfoot and the pattern of rearfoot motion during walking. Journal of Orthopaedic and Sports Physical Therapy 23:370–4.

McGeer T (1990) Passive dynamic walking. International Journal of Robotics Research 9:68–82.

McPoil TG, Hunt GC (1995) Evaluation and management of foot and ankle disorders: present problems and future directions. Journal of Orthopaedic and Sports Physical Therapy 21:381–8.

Menz HB, Lord SR (2005) Gait instability in older people with hallux valgus. Foot & Ankle International 26:483–9.

Nester CJ (1997) Rearfoot complex: a review of its interdependent components, axis orientation and functional model. The Foot 7:86–96.

Nigg BM (2001) The role of impact forces and foot pronation:a new paradigm. Clinical Journal of Sports Medicine 11:2–9.

Nigg BM, Denoth J, Neukomm PA (1981) Quantifying the load on the human body: problems and some possible solutions. In: Morecki A, Fidelus K, Kedzior K (eds) Biomechanics VIIB. University Park Press, Baltimore, pp 88–98.

Norkin CC, Levangie PK (1992) Joint structure and function: a comprehensive analysis. FA Davis Company, Philadelphia.

Orendurff MS, Segal AD, Klute GK, et al (2004) The effect of walking speed on center of mass displacement. Journal of Rehabilitation Research & Development 41(6A):829–34.

Ortega JD, Farley CT (2005) Minimizing center of mass vertical movement increases metabolic cost in walking. Journal of Applied Physiology 99:2099–2107.

Payne CB (1998) The past, present, and future of podiatric biomechanics. Journal of the American Podiatric Medical Association 88:53–63.

Peat M (1986) Functional anatomy of the shoulder. Physical Therapy 66:1855–65.

Perlman M, Leveille D, DeLeonibus J, et al (1987) Inversion lateral ankle trauma: differential diagnosis, review of the literature, and prospective study. Journal of Foot Surgery 26:95–135.

Perry J (1992) Gait analysis: normal and pathological function. SLACK Inc, Thorofare, NJ.

Peters JW, Trevino SG, Renström PA (1991) Chronic lateral ankle instability. Foot and Ankle 12:182–91.

Pierrynowski MR, Smith SB (1996) Rearfoot inversion/eversion during gait relative to the subtalar joint neutral position. Foot Ankle International 17:406–12.

Praet SFE, Louwerens J-WK (2003) The influence of shoe design on plantar pressures in neuropathic feet. Diabetes Care 26:441–5.

Root ML, Orien WP, Weed JH (1977) Normal and abnormal function of the foot. Clinical Biomechanics Corporation, Los Angeles.

Salathe EP, Arangio GA (2002) A biomechanical model of the foot: the role of muscles, tendons, and ligaments. Journal of Biomechanical Engineering 124:281–7.

Saunders M, Inman VT, Eberhart HD (1953) The major determinants in normal and pathological gait. Journal of Bone and Joint Surgery 35A:543–58.

Schaff PS, Cavanagh PR (1990) Shoes for the insensitive foot: the effect of a "rocker bottom" shoe modification on plantar pressure distribution. Foot Ankle 11:129–40.

Scott SH, Winter DA (1988) Internal forces at chronic running injury sites. Medicine and Science in Sports and Exercise 22:357–69.

Singh AK, Starkweather KD, Hollister AM, et al (1992) Kinematics of the ankle: a hinge axis model. Foot Ankle 13:439–46.

Sobel E, Caselli MA, Positano RG (1999) Orthotic therapy in the treatment of heel, Achilles tendon, and ankle injuries. In: Ranawat CS, Positano RG (eds) Disorders of the heel, rearfoot, and ankle. Churchill Livingstone, Edinburgh, pp 464–79.

Sutherland DH (2001) The evolution of clinical gait analysis part 1: kinesiological EMG. Gait and Posture 14:61–70.

Tanaka Y, Takakura Y, Takaoka T, et al (1997) Radiographic analysis of hallux valgus in women on weightbearing and nonweightbearing. Clinical Orthopaedics and Related Research 336:186–94.

Thomas S, Barrington R (2003) Hallux valgus. Current Orthopaedics 17:299–307.

Valmassy RL (1996) Clinical biomechanics of the lower extremity. CV Mosby, St Louis.

Voloshin A (2000) The influence of walking speed on dynamic loading on the human musculoskeletal system. Medicine and Science in Sports and Exercise 32:1156–9.

Wikipedia (2007) Available online at: http://en.wikipedia.org/wiki/Orthosis

Williams PL, Bannister LH, Berry MM, et al, eds (1995) Gray's anatomy. Longman, Edinburgh.

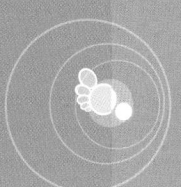

Glossary

Acceleration: the rate of change of speed, i.e. change in speed divided by change in time.

Actin: a protein that largely comprises the thin filaments within a sarcomere.

Action potential: depolarization of the cell membrane of a neurone resulting in the transmission of an impulse.

Active insufficiency: the lower limit, minimum length, of the working range of a muscle.

Anatomical position: reference body posture for descriptive purposes. The body is upright with the arms by the sides, palms of the hands facing forward, and feet slightly turned out.

Angular motion: angular motion, also referred to as rotation, occurs when a body or part of a body moves in a circle or part of a circle about a particular line in space, referred to as the axis of rotation, such that all parts of the body move through the same angle in the same direction in the same time.

Ankle joint: the joint between the distal tibia, distal fibula and the superior, lateral and medial aspects of the talus.

Antagonistic pair of muscles: the muscles or groups of muscles that move a joint in opposite directions about a particular axis of rotation.

Aponeurosis: a broad sheet of regular collagenous connective tissue that attaches a muscle to the skeletal system.

Apophyseal plate: a region of a bone that separates an apophysis from the rest of the bone prior to maturity; an apophyseal plate is responsible for growth of the bone adjacent to the non-apophyseal side of the plate.

Apophysis: a tuberosity separated from the rest of a bone prior to maturity by an apophyseal plate.

Appendicular skeleton: the bones of the upper and lower limbs.

Appositional growth: the type of growth in which new tissue is laid down on the surface of existing tissue.

Articular cartilage: the layer of hyaline cartilage that covers each articular surface of a bone (in a synovial joint).

Articular disc: a piece of fibrocartilage that helps to increase congruence in some synovial joints; the fibrocartilage may be in the form of a ring that tapers from the outside towards the centre.

Articular system: all of the joints of the body.

Aspect: the appearance of a particular bone (or any other part of the body) from a particular viewpoint.

Association neurone: a neurone that only has synapses at its nerve endings; also called an interneurone.

Attraction force: a force exerted between two bodies that tends to make the bodies move towards each other if they are apart and to maintain contact with each other after contact is made.

Autonomic nervous system: one of two functional divisions of the nervous system; it is concerned with the control of the visceral muscles, the heart and the exocrine and endocrine glands.

Axial skeleton: the bones of the skull, vertebral column and ribcage.

Axon: a nerve fibre that conducts impulses away from the cell body.

Biarticular muscle: a skeletal muscle that spans more than one joint.

Biomechanics: the study of the forces that act on and within living organisms and the effect of the forces on the size, shape, structure and movement of the organisms.

Bursa: a sac of synovial membrane containing synovial fluid interposed between structures that move relative to each other to minimize friction between them.

Bursitis: a painful condition resulting from inflammation of a bursa.

Cancellous bone: bone in which the osteones are loosely packed with spaces between them filled with red marrow; the osteones are usually arranged in the form of trabeculae.

Capsular ligament: a distinct thickening in part of the joint capsule that provides additional strength in one direction.

Cardiovascular system: the heart and blood vessels.

Cartilaginous joint: a joint in which the opposed surfaces of the bones are united by cartilage.

Cell: the fundamental structural and functional unit of all living organisms.

Cellular differentiation: the formation of four types of cell called tissues: epithelia, nerve, muscle and connective.

Central nervous system: one of two structural divisions of the nervous system consisting of the brain and spinal cord.

Centre of gravity: the point at which the whole weight of an object can be considered to act.

Centre of pressure: the point at which the ground reaction force can be considered to act.

Cerebrospinal nervous system: one of two functional divisions of the nervous system; it is concerned with consciousness and mental activities and control of skeletal muscle.

Chondral modeling: modeling of endochondral regions of the skeleton.

Collagen: one component of the matrix of ordinary connective tissue; collagen fibers break after being stretched approximately 10% of their rest length.

Compact bone: bone in which the osteones are closely packed with little or no space between them.

Composite material: a material that is stronger than any of the separate substances from which it is made.

Concentric contraction: a contraction in which the muscle shortens.

Congruence: the area over which the joint reaction force is transmitted in a synovial joint; in any particular joint position, the greater the contact area between the articular surfaces, the greater the congruence, and vice versa.

Contact force: a force exerted on a body due to physical contact with its environment.

Contractility: the ability of muscle tissue to create a pulling force.

Coronal plane: vertical plane perpendicular to the median plane.

Cranial nerves: 12 pairs of peripheral nerves that arise from the base of the brain.

Damping: a low level of resilience; a damping material returns very little energy during unloading compared to the amount of energy absorbed during loading.

Degrees of freedom: the linear and angular directions of movement considered normal for a joint with respect to reference anteroposterior, transverse and vertical axes, providing six possible degrees of freedom, three linear directions (along the axes) and three angular directions (around the axes).

Dendrite: a nerve fibre that conducts impulses towards the cell body.

Density: the amount of mass per unit volume.

Diaphysis: the shaft of a bone.

Digestive system: the alimentary canal and associated structures that break down food and eliminate solid waste.

Distance: (a) the length of the line between two points in three-dimensional space; (b) the length of the path followed by a body as it moves from one position to another in three-dimensional space.

Dorsiflexion of the ankle joint (true flexion of the ankle): sagittal plane motion of the foot about the ankle joint in which the dorsal surface of the foot is drawn closer to the shin.

Dynamics: the subdiscipline of mechanics that is concerned with the study of bodies under the action of unbalanced forces.

Eccentric contraction: a contraction during which the muscle lengthens.

Elastic tissue: regular or irregular elastic tissue.

Elasticity: the ability of a material to deform immediately in response to loading and, following unloading, to immediately restore its original size and shape.

Elastin: one component of the matrix of ordinary connective tissue; elastin fibers break after being stretched approximately 200% of their rest length.

End organs: sensory and motor nerve endings.

Endochondral ossification: another name for intracartilagenous ossification.

Endocrine gland: a ductless gland that discharges secretions directly into the blood stream.

Endocrine system: the glands that secrete hormones which regulate and coordinate the various body functions in association with the nervous system.

Energy: the capacity to do work.

Epiphyseal plate: the region of a bone between the epiphysis and the diaphysis that is responsible for growth in length of the bone by interstitial growth.

Epiphysis: end of a bone separated from the diaphysis by an epiphyseal plate prior to maturity.

Equilibrium: with regard to linear motion, an object is in equilibrium when the resultant force acting on the object is zero, i.e. when the object is at rest or moving with constant linear velocity. With regard to angular motion, an object is in equilibrium when the resultant moment acting on the object is zero, i.e. when the object is at rest or moving with constant angular momentum.

Eukaryotic cell: a cell that has a nucleus and organelles. Organisms made up of one (unicellular) or more (multicellular) eukaryotic cells are called eukaryotes. The human body is a eukaryote.

Exocrine gland: a gland with a duct that discharges secretions onto a surface.

External force: in biomechanical analysis of human movement, the forces that act on the body from external sources, such as body weight, ground reaction force, water resistance and air resistance, are referred to as external forces.

Extracapsular ligament: a non-capsular ligament outside the joint cavity.

Fascia: any type of ordinary connective tissue in the form of a sheet.

Fibrous joint: a joint in which the opposed surfaces of the bones are united by fibrous tissue.

Fibrous tissue: regular or irregular collagenous connective tissue.

Fluid mechanics: the study of the forces that act on bodies in fluids and the effects of the forces on the movement of the bodies.

Force: that which alters or tends to alter a body's state of rest or type of movement.

Force-velocity relationship: the relationship between speed of shortening or speed of lengthening and tension in a muscle.

Free body diagram: a diagram showing all of the forces acting on an object.

Friction: when one object moves or tends to move across the surface of another, there will be a force parallel to the surfaces in contact that will oppose the movement or tendency to move; this force is called friction.

Glial cell: a specialized connective tissue cell found only in the nervous system.

Gravitational potential energy: the energy possessed by an object by virtue of its height above any particular reference position, usually ground level.

Gravity: the acceleration experienced by a body due to the force of attraction (weight of the body) exerted on the body by the earth.

Ground reaction force: the force exerted by the ground on the body.

Ground substance: the nonfibrous component of the matrix of ordinary connective tissue.

Hysteresis loop: the loop on a load-deformation curve defined by the loading and unloading phases; the loop defines the amount of strain energy dissipated between the end of the loading phase and the end of the unloading phase.

Impact: a collision, usually of short duration, between two bodies.

Impact force peak: the peak force during the passive phase of impact of the human body with another body or object.

Impulse of a force: the product of the magnitude of a force and the duration of the force.

Inertia: the resistance/reluctance of a body to start moving if it is stationary or to change its speed and/or direction if it is already moving.

Integumentary system: the external covering of the body, i.e. the skin and associated structures such as finger and toe nails.

Internal force: in biomechanical analysis of human movement, muscle forces and the forces acting on ligaments and bones are referred to as internal forces.

Interstitial growth: the type of growth in which new tissue is produced from within the mass of existing tissue.

Intracapsular ligament: a non-capsular ligament inside the joint cavity.

Intracartilagenous ossification: ossification of hyaline cartilage.

Intramembranous ossification: ossification of fibrous membranes.

Isometric contraction: a contraction during which the length of the muscle does not change.

Isotonic contraction: a contraction involving a change in the length of the muscle (concentric contraction, eccentric contraction).

Joint: a region where two or more bones are connected.

Joint complex: a group of joints with a relatively high degree of functional interdependence.

Joint reaction force: the equal and opposite force exerted on articular surfaces in contact with each other in a joint.

joule (J): the unit of work in the SI system of units. 1 J is the work done by a force of 1 newton (N) when it moves its point of application a distance of 1 metre (m) in the direction of the force.

Kinematics: the branch of dynamics that describes the movement of bodies in relation to space and time.

Kinesthetic sense: awareness of body position and body movement.

Kinetic energy: the energy possessed by an object by virtue of its speed.

Kinetics: the branch of dynamics that describes the forces acting on bodies, i.e. the cause of the observed kinematics.

Length-tension relationship: the relationship between length and tension in a muscle.

Lever: a rigid or quasi-rigid object that can be made to rotate about a fulcrum in order to exert a force on another object.

Linear momentum: the product of mass and linear velocity.

Linear motion: linear motion, also referred to as translation, occurs when all parts of a body move the same distance in the same direction in the same time.

Load: a load is any force or combination of forces applied to an object.

Lymphatic system: the system of vessels and ducts that drains extracellular fluid and returns it to the blood.

Mass: the amount of matter (physical substance) that comprises a body. The mass of a body is a measure of the inertia of the body.

Matrix: the non-cellular component of connective tissue.

Mechanical advantage: a measure of the efficiency of a lever system; the ratio of the resistance force to the effort force or the ratio of the moment arm of the effort force to the moment arm of the resistance force.

Mechanical efficiency of the human body: the ratio of the mechanical work done by the body during a particular period of time to the corresponding metabolic energy expenditure, usually expressed as a percentage.

Mechanical energy: the energy of an object by virtue of its height above a particular reference point (gravitational potential energy) and its speed (kinetic energy).

Mechanical power: the rate at which energy is transformed in the form of work.

Mechanics: the study of the forces that act on bodies and the effects of the forces on the size, shape, structure and movement of the bodies.

Mechanics of materials: the subdiscipline of mechanics that is concerned with the study of the mechanical properties of materials.

Median plane: the vertical plane that divides the body down the middle into more-or-less symmetrical left and right portions.

Meniscus: a C-shaped wedge of fibrocartilage that helps to increase congruence in the tibiofemoral joint; there are normally two menisci in each tibiofemoral joint.

MET (metabolic equivalent:) a method of comparing energy expenditure in different activities based on resting metabolic rate. One MET (resting metabolic rate) varies slightly among individuals, but is approximately $3.5\,ml.kg^{-1}.min^{-1}$ of oxygen.

Metabolic energy expenditure: the energy cost of metabolism, usually expressed as energy expenditure.

Metabolic rate: the rate of metabolic energy expenditure, usually expressed in relation to body mass and time as, for example, energy expenditure per kilogram of body mass per minute.

Metabolism: all of the chemical reactions that take place in the body.

Metaphysis: the region of a bone where the epiphysis joins the diaphysis; in a growing bone this corresponds to the calcified layer of the epiphyseal plate together with the interdigitating bone.

Midtarsal joint (transverse tarsal joint): the calcaneocuboid and talonavicular joints.

Modeling: changes in the expression of the skeletal genotype that occur as a result of environmental influences.

Moment of a force: the product of the magnitude of the force and the perpendicular distance between the axis of rotation and the line of action of the force.

Motor nerve ending: a nerve ending in contact with a specialized effector organ such as a motor end plate in muscle.

Motor neurone: a neurone that conducts impulses away from the brain; also called an efferent neurone.

Motor unit: the functional unit of skeletal muscle consisting of an alpha motor neurone together with all the terminal branches of the axon and the muscle fibres that they innervate.

Multicellular organism: an organism that consists of many cells.

Muscle fibre: a muscle cell.

Muscle spindle: a tension receptor in skeletal muscle.

Muscle tone: the resting level of tension in muscle; the level of muscle tone in the extrafusal fibres is determined by the level of muscle tone in the intrafusal fibres.

Muscular system: the skeletal muscles.

Musculoskeletal system: the skeletal system and the muscular system.

Musculoskeletal system function: the mechanical function of the musculoskeletal system is to generate and transmit internal forces to counteract the effects of gravity and create the ground reaction forces (and propulsion forces in water) necessary to maintain upright posture, transport the body and manipulate objects, often simultaneously.

Musculotendinous unit: a skeletal muscle and its tendons and/or aponeuroses.

Myelinated nerve fibre: a nerve fibre that has a myelin sheath.

Myofibril: a chain of sarcomeres.

Myosin: a protein that largely comprises the thick filaments within a sarcomere.

Nerve fibres: the processes that extend from the cell body of a neurone.

Nervous system: the nerves, organized into central (brain and spinal cord) and peripheral (spinal nerves) components.

Neurone: a nerve cell.

newton: the unit of force in the SI system of units. A newton is defined as the force acting on a mass of 1 kg that accelerates it at $1 \, m/s^2$.

Newton's first law of motion: the resultant force acting on a body at rest or moving with uniform linear velocity is zero and the body will remain at rest or continue to move with uniform linear velocity unless the resultant force acting on it becomes greater than zero.

Newton's law of gravitation: every body attracts every other body with a force that varies directly with the product of the masses of the two bodies and inversely with the square of the distance between them.

Newton's second law of motion: when a force (resultant force greater than zero) acts on a body, the change in momentum experienced by the body takes place in the direction of the force and is directly proportional to the magnitude of the force and the length of time that the force acts.

Newton's third law of motion: whenever one body A exerts a force on another body B, body B simultaneously exerts an equal and opposite force on body A.

Non-capsular ligament: a distinct band separate from the joint capsule or only partially attached to it.

Non-myelinated nerve fibre: a nerve fibre that does not have a myelin sheath.

Non-pennate muscle: a muscle in which the fibres run in line with the line of pull of the muscle.

Organ: a combination of different tissues designed to carry out a specific bodily function.

Osteone: a column of bone consisting of three to nine concentric layers of bone surrounding a haversian canal.

Osteopenia: bone density below the normal level for the age and sex of the individual.

Osteoporosis: loss of bone mass and elasticity to the extent that bones are no longer able to withstand the loads imposed by normal, habitual activity, resulting in a high susceptibility to fracture.

Passive insufficiency: the upper limit, maximum length, of the working range of a muscle.

Pennate muscle: a muscle in which the fibres run obliquely with respect to the origin and insertion so that the line of pull of the fibres is oblique to the line of pull of the muscle.

Periosteum: the layer of fibrous tissue that covers the non-articular surfaces of a bone; the periosteum is responsible for growth in girth of a bone by appositional growth.

Peripheral nervous system: one of two structural divisions of the nervous system; it consists of 43 pairs of nerves (bundles of nerve fibres), which arise from the base of the brain and the spinal cord.

Plantar flexion of the ankle (extension of the ankle): sagittal plane motion of the foot about the ankle joint in which the dorsal surface of the foot is moved away from the shin (pointing the toes).

Plexus: network of spinal nerves that innervate a specific region of the body.

Podiatric biomechanics: the branch of biomechanics concerned with the study of the forces that act on and within the foot and lower leg during weightbearing activity and the effects of the forces on the structure and function of the foot.

Porosity: the proportion of non-bone tissue in a bone or region of a bone.

Potential energy: stored energy.

Power: the rate of transformation of energy from one form to another.

Prokaryotic cell: a cell that does not have a nucleus or organelles. Organisms made up of prokaryotic cells are called prokaryotes. Most prokaryotes are unicellular, but some prokaryotes have multicellular stages in their life cycles. Prokaryotes include bacteria. Prokaryotic cells were the first type of cell to evolve.

Pronation of the foot: simultaneous abduction, dorsiflexion and eversion of the foot.

Proprioception: the sense of the position of a joint and movement of a joint; proprioception is part of kinesthetic sense.

Qualitative analysis of movement: analysis based on observation.

Quantitative analysis of movement: analysis based on measurements of the variables that determine performance.

Rearfoot complex: the ankle, subtalar and midtarsal joints. The rearfoot complex facilitates pronation and supination.

Reflex action: an involuntary movement that provides protection by rapidly removing part of the body from a potential source of danger without conscious effort.

Remodelling: coordination of osteoblastic and osteoclastic activity resulting in changes in external form and internal architecture of the bones, including repair of bones.

Reproductive system: the ovaries and associated structures (female) and testes and associated structures (male) that enable the body to produce offspring.

Resilience: the amount of energy returned during unloading as a percentage of the amount of energy absorbed during loading.

Resolution of a vector: the process of replacing a vector by two or more component vectors.

Respiratory system: the lungs and associated passageways.

Response to loading: the immediate changes in stress and strain experienced by musculoskeletal components following a change in loading.

Running economy: the rate of energy expenditure for a given submaximal running speed.

Sagittal plane: any plane parallel to the median plane.

Sarcomere: the basic structural unit of a muscle fibre that contains the contractile apparatus.

Scalar quantity: a quantity that can be completely specified by its magnitude (size).

Schwann cell: the only type of glial cell in the peripheral nervous system. Schwann cells produce a myelin sheath around many nerve fibres. Myelinated nerve fibres conduct nerve impulses much faster than non-myelinated nerve fibres.

Sensory nerve ending: a nerve ending in contact with a specialized receptor organ such as a pain receptor.

Sensory neurone: a neurone that conducts impulses towards the brain; also called an afferent neurone.

Skeletal genotype: the process of genetically programmed change in the external form (size and shape) and internal architecture of the bones.

Skeletal system: the skeleton and joint support structures.

Sliding filament theory: the generally agreed mechanism of muscle contraction involving coupling, decoupling and sliding between actin and myosin filaments.

Speed: the rate of change of position, i.e. distance moved divided by the time taken to move the distance.

Spinal nerves: 31 pairs of peripheral nerves that arise from the spinal cord.

Stability: with respect to a particular base of support, an object is stable when the line of action of its weight intersects the plane of the base of support and unstable when it does not.

Statics: the subdiscipline of mechanics that is concerned with the study of bodies under the action of balanced forces.

Stiffness: the resistance of a material to deformation.

Strain: in mechanics, the deformation of an object that occurs in response to a load is referred to as strain.

Strain energy: the energy stored in an object as a result of being strained.

Strength of a material: the amount of force required to break the material.

Stress: the resistance of the intermolecular bonds of an object to the strain caused by a load.

Stretch reflex: the involuntary contraction of a muscle in response to a sudden unexpected increase in its length; the reflex is brought about by the spindle afferent to muscle efferent loop.

Structural adaptation: the permanent change in the external form or internal architecture of musculoskeletal components that occurs as a result of changes in the time-averaged loads exerted on them.

Subtalar joint: the joint between the talus and calcaneus.

Supination of the foot: simultaneous adduction, plantar flexion and inversion of the foot.

Symphysis: a cartilaginous joint in which the opposed surfaces are united by a combination of hyaline cartilage and fibrocartilage; a layer of hyaline cartilage covers each of the articular surfaces, and sandwiched between the layers of hyaline cartilage is a relatively thick pad of fibrocartilage.

Synapse: where a nerve ending is in contact with another neurone.

Synchondrosis: a cartilaginous joint in which the opposed surfaces are united by hyaline cartilage.

Syndesmosis: another name for a fibrous joint.

Synovial bursa: a closed sac comprised of synovial membrane containing synovial fluid that is interposed between different tissues that slide on each other to prevent friction.

Synovial joint: a joint in which the opposed (articular) surfaces are not attached to each other, but are held in contact with each other by ligaments and a joint capsule.

Synovial membrane: a highly specialized form of connective tissue that lines the capsule in a synovial joint. The synovial membrane produces synovial fluid that lubricates and nourishes the articular cartilage.

Synovial sheath: a closed flattened sac comprised of synovial membrane containing a capillary film of synovial fluid. A synovial sheath forms a protective sleeve around a tendon or ligament to prevent friction with adjacent tissue.

System: a combination of different organs working together to carry out a particular bodily function.

Tendon: a cord-shaped or narrow band of regular collagenous connective tissue that attaches a muscle to the skeletal system.

Tenosynovitis: a painful condition resulting from inflammation of a synovial sheath.

Tissue: a group of cells having the same specialized structure, enabling them to perform a particular function in the body. There are four tissues: epithelia, nerve, muscle and connective.

Transverse plane: a horizontal plane, perpendicular to both the median and coronal planes, that divides the body into upper and lower portions.

Trigonometry: the branch of mathematics that deals with the relationships between the lengths of the sides and the sizes of the angles in a triangle.

Unicellular organism: an organism that consists of a single cell.

Urinary system: the kidneys, bladder and associated structures that eliminate nitrogenous waste as urine.

Vector chain: the method of determining the resultant of a number of component vectors by joining the component vectors together in a chain.

Vector quantity: a quantity that requires specification in both magnitude and direction.

Viscoelasticity: the ability of a material to gradually deform in response to loading and, following unloading, to gradually restore its original size and shape.

Viscosity: the resistance of a fluid to flowing.

Volume: the amount of space that a mass occupies.

watt: work rate of 1 joule per second (J/s).

Weight of a body: the force of attraction exerted on a body by the earth.

Wolff's law: bone adapts its external form and internal architecture to the time-averaged load exerted on it in an ordered and predictable manner to provide optimal strength with minimal bone mass.

Work: a force does work when it moves its point of application in the direction of the force and the amount of work done is defined as the product of the force and the distance moved by the point of application of the force.

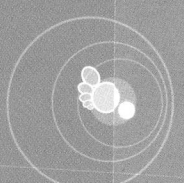

Index

Note to index: for definitions, consult the glossary